A DICTIONARY OF

BOTANICAL TERMS

MAVEN BOOKS

A DICTIONARY OF
BOTANICAL TERMS

A.A. Crozier

MAVEN BOOKS

Chennai New Delhi Tirunelveli

MAVEN BOOKS

An Imprint of MJP Publishers

ISBN 978-93-87488-45-8

MAVEN BOOKS
No. 44, Nallathambi Street,
Triplicane, Chennai 600 005

Printed and bound in India

MJP 314

Publisher: **C. Janarthanan**

PUBLISHER'S NOTE

The legacy of a country is in its varied cultural heritage, historical literature, developments in the field of economy and science. The top nations in the world are competing in the field of science, economy and literature. This vast legacy has to be conserved and documented so that it can be bestowed to the future generation. The knowledge of this legacy is slowly getting perished in the present generation due to lack of documentation.

Keeping this in mind, the concern with retrospective acquiring of rare books has been accented recently by the burgeoning reprint industry. Maven Books is gratified to retrieve the rare collections with a view to bring back those books that were landmarks in their time.

In this effort, a series of rare books would be republished under the banner, "Maven Books". The books in the reprint series have been carefully selected for their contemporary usefulness as well as their historical importance within the intellectual. We reconstruct the book with slight enhancements made for better presentation, without affecting the contents of the original edition.

Most of the works selected for republishing covers a huge range of subjects, from history to anthropology. We believe this reprint edition will be a service to the numerous researchers and practitioners active in this fascinating field. We allow readers to experience the wonder of peering into a scholarly work of the highest order and seminal significance.

MAVEN BOOKS

PREFACE.

It is now nearly twenty years since an English dictionary of botanical terms was published, and the development of botany during that time has brought into use many new terms and led to the abandonment of some of those which were formerly employed. The present work, though larger than any of its predecessors, is therefore wanting in certain terms which are found in the earlier botanical dictionaries. A few obsolete terms, however, which occur in standard botanical literature are retained and marked as such. The scope of this dictionary is nearly the same as that of its predecessors, except that it is broader on the side of agriculture and horticulture, as it aims to include all technical terms applied to plants both by botanists and others. Very many of the newer terms are from the German botanists, to whom we owe by far the greater part of modern structural and physiological botany. These terms, particularly, include an unfortunately large number of synonyms, which seem to be a necessary result of active research by independent workers in the same fields. With few exceptions, no definition is repeated, all approved synonyms being brought together under one term. Where a choice existed this has permitted the preferable term alone to be defined. No obsolete or wholly undesirable synonyms, however, accompany the definitions, and such of these as occur in alphabetical order have usually been admitted only to refer the reader to the proper term. The cross-references freely given to terms of similar and opposite meanings will be found useful in broadening the scope of the definitions.

I desire here to express my gratitude to various botanical friends for their contributions to this work, particularly to Dr. W. J. Beal, of the Michigan Agricultural College, who has revised the entire manuscript and improved many of the definitions; to Professor V. M. Spalding, of the University of Michigan, who has revised and corrected the terms relating to fungi; to Mr. F. C. Newcombe for aid upon the terms applied in karyokinesis; and to Professor L. H. Bailey, of Cornell University, for advance proof-sheets of his glossary of horticultural terms.

The marking of the pronunciation was begun by Mr. B. Pickman Mann, of Washington, D. C., and completed by Mr. F. P. Jordan, of the University of Michigan. In their work Webster's dictionary was mainly followed as authority.

ANN ARBOR, MICH January, 1892.

SIGNS AND ABBREVIATIONS.

①....annual.
②....biennial.
♃....perennial.
♂....staminate.
♀....pistillate.
☿....perfect.
∞....numerous; more than twenty when applied to stamens.
?.....doubt.
!.....certainty. Indicates that the author has seen the specimen in question.
n. sp. new species. Used only with the first printed description.
Eu...the Greek word "well," sometimes written after a species to indicate that it is certainly a well-defined species, not a variety.

MEASUREMENT.

°.....foot.
′.....inch.
″.....line ($\frac{1}{12}$ inch).
c.m...centimeter.
c.c...cubic centimeter.
m.m. millimeter.
μ....*mu* (the Greek letter *m*), micromillimeter, $\frac{1}{1000}$ of a millimeter, the unit of microscopic measurement. It is about .000039 of an inch.
×....used to express magnification. Thus, × 150 indicates a magnification of 150 diameters. The improper fraction $\frac{150}{1}$ indicates the same, but this form is little used.

A DICTIONARY

OF

BOTANICAL TERMS.

A, prefixed to words of Greek origin often signifies absence, as *ăpĕt'ălous*, without petals. Before a vowel it is changed to AN, as *ănăn'therous*, without anthers.

ĂBĂX'ĬAL, out of the axis. See EXCENTRIC.

ĂBBRĔ'VIĀTED, shorter than an adjoining organ, or one with which it is compared.

ĂBĔR'RANT, differing widely from the usual form or structure; as: the Fumariaceæ have been regarded as an *aberrant* group of Papaveraceæ.

ĂBĬŌGĔN'ESĬS, see SPONTANEOUS GENERATION.

ĂBJĔC'TION, throwing off with force, as spores or seeds.

ĂBJOINT', to separate at a joint.

ĂBJŬNC'TION, the separation of one part from another; delimitation. Compare ABSTRICTION.

ĂBNÔR'MAL, differing from the usual structure or condition.

ĂBŌRĬG'ĬNAL, see INDIGENOUS.

ĂBÔR'TION, the non-formation or imperfect formation of an organ, so that it does not perform its usual function. Compare SUPPRESSION, DEGENERATION, ATROPHY.

ĂBŎR'TĬVE, imperfect, or wanting. See OBSOLETE.

ĂBRŬPT', terminating suddenly, as a leaf which ends without a tapering extremity, a compound leaf without a terminal leaflet, or a stem which is sharply bent.

ĂBRŬPT'LY̆ ĂCŪ'MĬNĀTE, having a broad extremity from which a point arises.

ĂBRŬPT'LY̆ PĬN'NĀTE, pinnate without a terminal leaflet, and usually with an even number of leaflets. There may be a terminal point or tendril; even pinnate; equally pinnate. Compare IMPARIPINNATE.

ĂBSÇĬȘ'ȘION, cutting off sharply, as the separation of the frustules of diatoms. Compare ABSTRICTION.

ĂB'SÇĬSS LĀY'ER, a layer of cells formed by renewed cell-division in the base of the leaf-stalk in autumn, the formation of which permits the fall of the leaf.

ĂBSŎRP′TION, the method by which fluids enter the plant. See OSMOSE and IMBIBITION.

ĂBSTRĬC′TION, the separation of one part from another by constriction and the formation of a septum. Applied especially to the separation of spores from their hyphæ. Compare ABSCISSION and ABJUNCTION.

ĂCĂL′ỸÇĬNE, without calyx; acalycinous.

ĂCĂLỸÇ′ĬNOŬS, see ACALYCINE.

ĂCĂN′THĂ (pl. Ăcăn′thæ), a thorn, spine, or prickle.

ĂCĂNTHĀ′ÇEOŬS, having thorns, spines, or prickles.

ĂCĂNTHŌCÄR′POŬS, having the fruit furnished with spines or prickles.

ĂCĂNTHŎC′LĂDOŬS, having spiny branches.

ĂCĂNTHŎPH′ŎROŬS, bearing prickles, spines, or thorns.

ĂCĂNTHŎP′ŎDOŬS, having the petiole or pedicel furnished with spines or prickles.

ĂCÄR′POŬS, not producing fruit.

ĂCAULĔS′ÇENÇE, an abnormal suppression of the stem.

ĂCAULĔS′ÇENT, having only a very short aërial leaf-bearing stem, apparently none, as in the dandelion; acauline; acaulose; acaulous.

ĂCAU′LĬNE, see ACAULESCENT.

ĂCAU′LĬS, see ACAULESCENT.

ĂCAU′LŎSE, see ACAULESCENT.

ĂCAU′LOŬS, see ACAULESCENT.

ĂCÇES′SŎRỸ, additional to the usual number, or accompanying something else; supernumerary. Said of the additional buds when more than one occur in an axil; applied to the border of the apothecium in lichens when of a different substance or color from the disk.

ĂCÇES′SŎRỸ BUDS, also called secondary buds and supernumerary buds; the additional buds when more than one occur in or near the axil, as in the butternut.

ĂCÇES′SŎRỸ CELL, the sister-cell of a guard-cell; subsidiary cell.

ĂCÇES′SŎRỸ FRUIT, one in which some additional part or parts is associated with the matured ovary, as in the wintergreen, *Gaultheria procumbens;* anthocarpous fruit. The term is not applied to fruits with an adherent calyx-tube, as the apple and currant. Compare AGGREGATE FRUIT and COLLECTIVE FRUIT.

ĂCÇES′SŎRỸ GŎNĬD′ĬĂ, gonidial formations in some species of Mucorini in addition to the typical kind.

ĂCÇĬDĔN′TĂL, see ADVENTITIOUS.

ĂCCŎMMODĀ′TION, see ADAPTATION.

ĂCCRĔS′ÇENT, increasing in size after flowering;—applied to parts accessory to the fruit, as the calyx of Physalis.

ĂCCRĒTE′, grown together in any manner.

ĂCCRĒ′TION, the growing of one thing to another; growth by addition of particles to the outside.

ĂCCŬM′BENT, lying upon or against another body, as the edge of the cotyledons against the caulicle in some Cruciferæ. Compare INCUMBENT.

ĂCĔPH′ĂLOŬS, headless; said of an ovary when the style is attached to some other part than the summit.

ĂÇERŌSE′, narrow, stiff, and

pointed, like the leaves of the fir. Intermediate in form between Acicular and Subulate.

ĂÇĔR'VĀTE, growing in heaps or clusters; coacervate.

ĂÇĔR'VŪLŬS (pl. **Ăçĕr'vūlī**), a small heap, as of spores.

ĂCETĂBŪ'LĬFŎRM, deep saucer-shaped, with a broad concave bottom and nearly upright sides. Compare COTYLIFORM.

ĂÇETĀ'RĬOŬS, used for salads.

ĂÇ'ETŌSE, sour; acid.

ĂCHÆ'NĬŬM, see ACHENIUM.

ĂCHÆ'NŎCĂRP, any simple dry indehiscent fruit. (Masters.)

ĂCHEĪ'LĂRY̆, destitute of a labellum, as some orchids.

ĂCHĒNE', see ACHENIUM.

ĂCHĒ'NĬŬM (pl. **Ăchē'nĭă**), a small, dry, indehiscent, one-seeded seed-like fruit or carpel in which the covering does not adhere to the seed, as in the sunflower or buttercup; achænium; achene; akene. Compare CARYOPSIS, UTRICLE, CYPSELA.

ĂCHĒNŌ'DĬŬM, see CREMOCARP.

ĂCHLĂMY̆'DEOŬS, destitute of calyx and corolla.

ĂCHROMĂT'ĬC, not readily colored by the usual staining agents.

ĂCHRŌ'MĂTĬN, the basic substance of the nucleus, less highly colored by staining agents than the rest.

ĂÇĬC'ŪLĂ (pl. **Ăçĭc'ūlæ**), a needle-shaped spine, prickle, or other body.

ĂÇĬC'ŪLAR, slender needle- or bristle-shaped, as the leaves of most pines; aciculate; aciculiform; aciform. Compare ACEROSE.

ĂÇĬC'ŪLĀTE, see ACICULAR.

ĂÇĬC'ŪLĀTED, marked by fine grooves, as if scratched with a needle. Finer than striate, and not necessarily parallel.

ĂÇĬC'ŪLĬFÔRM, see ACICULAR.

Ă'ÇĬĒS, an edge.

ĂÇ'ĬFÔRM, see ACICULAR.

ĂÇĬNĀ'ÇEOŬS, full of kernels. (Rare.)

ĂÇĬNĂÇ'ĬFÔRM, scymitar-shaped;—having one edge thin and convex, and the other thick and straight or concave, like many bean-pods.

ĂÇ'ĬNE, one of the succulent carpels of a fruit like the raspberry; acinus. (Originally Acinus meant a bunch of grapes or similar fruit, as the currant, and later the term was applied to one of the berries of such a bunch and to the seeds of the grape.)

ĂÇĬNŌSE', consisting of granular bodies somewhat resembling grape-seed. More rarely, resembling a grape-berry or bunch of grapes.

ĂÇ'ĬNŬS (pl. **Ăç'inī**), see ACINE.

ĂCŎTY̆LĔD'ŎNOŬS, destitute of cotyledons, as Cuscuta.

ĂCRĂMPHĬB'RY̆OŬS, said of plants whose growth is not confined to the apex, i.e., which produce lateral as well as terminal buds. (Rare.) Compare. ACROGENOUS.

ĂCRŎB'RY̆OŬS, see ACROGENOUS.

ĂCRŎCĂR'POŬS, having terminal fructification. Used chiefly in mosses. Compare CLADOCARPOUS.

ĂCRŎĠ'ENOŬS, growing chiefly at the apex, as the stem of ferns; produced at the apex of a filament, as the spores of some fungi.

ĂCRŎĠ'Y̆NOŬS, having the stem terminated by archegonia or other female organs. Compare ANACROGYNOUS.

ĂCRŎP'ĔTĂL, produced in succession toward the apex; in the direction of the summit; basifugal; centripetal (when applied to the order of inflorescence only). Compare BASIPETAL.

ĂCRŎSCŎP'ĬC, facing the apex. Compare BASISCOPIC.

ĂC'RŎSPĪRE, an old name for the plumule of a grain in germination.

ĂC'RŎSPŌRE, a spore borne at the summit of a filament.

ĂCTĬNŎMÔR'PHĬC, regular and polysymmetrical, as the flowers of radish, i.e., capable of bisection in two or more planes into similar halves. Compare ZYGOMORPHIC.

ĂCTĬNŎMÔR'PHOŬS, see ACTINOMORPHIC.

ĂC'TĪVE, in a growing condition; opposed to Dormant.

ĂCŪ'LĔĀTE, having prickles, prickle-pointed; aculeated.

ĂCŪ'LEŎLĀTE, having small or few prickles.

ĂCŪ'LEŬS (pl. **Ăcū'leī**), see PRICKLE.

ĂCŪ'MĬNĀTE, ending in a prolonged tapering point.

ĂCŪ'MĬNŌSE, having a sharp or tapering point; somewhat acuminate. (Rare.)

ĂCŪMĬN'ŪLĀTE, having a small acuminate point.

ĂCŪTE', ending in a distinct angle, but not prolonged.

ĂÇȲC'LĬC, having all the organs of a flower in a continuous spiral. Compare HEMICYCLIC and CYCLIC.

ĂDĂPTA'TION, the modification of a plant to better fit it for surrounding conditions, or of an organ for a particular function; accommodation. Compare METAMORPHOSIS.

ĂDĔL'PHĬĂ, (pl. **Ădĕl'phĭæ**), a collection of stamens united by their filaments. Compare PHALANX.

ĂDĔL'PHOŬS, having the stamens united by their filaments in an adelphia or bundle.

ĂDĔN'ĬFÔRM, gland-shaped; adenoid.

ĂDĔN'ŎPHŎRE, a stalk supporting a gland.

ĂDENŎPH'ŎRǪŬS, gland-bearing.

ĂDENŎPHȲL'LOŬS, having on the leaves glands, or gland-like spots or tubercles.

ĂDENŎP'ŎDOŬS, having glands or gland-like tubercles on the petioles.

ĂDENŌSE', gland-like, or bearing glands; adenoid; adenous.

ĂD'ENOŬS, see ADENOSE.

ĂD'ESMȲ, the separation of parts usually joined. Compare CHORISIS, DIALYSIS, FISSION.

ĂDGLŪ'TĬNĀTE, see AGGLUTINATE.

ĂDHĔ'RENT, attached or growing to an organ or body of a different nature, as the calyx to the ovary. The parts may or may not have grown together from the first. Compare COHERENT and ADNATE.

ĂDHĒ'ȘION, the sticking or growing together of organs of a different nature, as a leaf to the stem. It implies a union of parts which in most other plants or in a younger state of the same plant are separate.

ĂDNĂS'ÇENT, growing to or upon something else; as, "moss is an *adnascent* plant."

ĂD'NĀTE, said of parts of a different nature which are grown together from the first, or over their whole surface or length. (Compare CONNATE.) Applied to an anther, it indicates that it is attached throughout its length to the upper or lower surface of the filament. Compare INNATE. See ADHERENT.

ĂDNĀ'TION, the same as adhesion, but implying an earlier or more complete union.

ĂDPRĔSS'ED, see APPRESSED.

ĂDSÇĔN'DĔNT, see ASCENDING.

ĂDSŨR'GENT, see ASCENDING.

ĂDŬN'COŬS, crooked, twisted, or hooked. (Rare.)

ĂDVENTĪ'TIŎUS, out of the usual place, as buds on a leaf or at a distance from a node; growing spontaneously out of its native locality, but not fully established; adventive.

ĂDVĔN'TĬVE, see ADVENTITIOUS.

ĂD'VĔRSE, (1) opposite (rare); (2) directed toward or facing the main axis or other object. Compare AVERSE.

ĂDVĔRSĬFŌ'LĬĀTE, see ADVERSIFOLIOUS.

ĂDVĔRSĬFŌ'LĬOŬS, having opposite leaves; adversifoliate.

ÆÇĬD'ĬŌSPŌRE, a spore produced in an æcidium.

ÆÇĬD'ĬŬM (pl. **Æçĭd'ĭā**), the cup-shaped spore-case with its contents in certain Uredineæ, in which the spores are produced serially from below; æcidium fruit.

ĀERĔN'CHȲMĀ, tissue homologous with cork, with thin cell-walls and large intercellular spaces, found in the stems of some marsh-plants, as *Lythrum salicaria*. It is supposed by Schenk, the author of the term, to serve for aeration, but by others it is regarded as floating tissue.

ĀĔ'RĬAL, growing in the air, not attached to the soil, as a parasite or epiphyte upon the top of some other plant.

ĀĔ'RĬĂL ROOTS, those appearing above ground, whether they afterwards enter the ground or not, as the clinging roots of the ivy and the brace-roots of Indian corn.

ĀERŎ'BĬĀ (sing. **Āerō'bĭum**), organisms which thrive only in the presence of air or free oxygen. Applied in bacteria. Compare ANAEROBIA.

ĀERĬŌBĬŌT'ĬC, thriving only in the presence of air.

Ā'ERŎÇȲST, an old name for air-bladder.

Ā'ERŎPHȲTE, see AIR-PLANT.

Ā'ERŌTRŎP'IC, seeking the air, as certain roots.

ÆRU'ĠĬNŌSE, clear light bluish green; verdigris-green.

ÆRU'ĠĬNOŬS, see ÆRUGINOSE.

ÆSTĪ VĂL, see ESTIVAL.

ÆSTĬVĀ'TION, the arrangement of the floral organs in the bud; præfloration.

ÆTHĀ'LĬŬM, a compound sporiferous body in Myxomycetes, formed from a large combination of plasmodia.

ĂFFĬN'ĬTȲ, true or near relationship. Compare ANALOGY and HOMOLOGY.

ĂGĂM'ĬC, destitute of sex. Formerly applied to all cryptogams.

ĂG'ĂMŎĠĔN'ESĬS, asexual reproduction of any kind. See PARTHENOGENSIS and APOGAMY.

ĂGĂM'ŎSPŌRE, a spore formed without fertilization; — a gonidium or asexually produced spore of any kind. (Rare.)

ĂG'ĂMOŬS, see Agamic.

ĂGGLŎM'ERĀTE, see Acglomerated.

ĂGGLŎM'ERĀTED, clustered or crowded together but not cohering, as the staminate flowers of pines.

ĂGGLŪ'TĬNĀTED, glued together. Compare Accrete.

ĂG'GREGĀTĔD, collected together but not cohering. About the same as Agglomerated.

ĂG'GREGĀTE FLOWERS, those with several in the same head, as in clover. The term excludes Compositæ.

ĂG'GREGĀTE FRUIT, one in which distinct carpels of a single flower are crowded on the receptacle into one mass, as in the raspberry and magnolia; syncarp. Compare Collective Fruit.

ĂGGREGĀ'TION, the condensation of the protoplasm of a living cell, or of some of the contained proteids under stimulation. first observed in the tentacle cells of Drosera and subsequently produced in the cells of other plants by means of various basic substances.

ĂGRĬCŬLT'ŪRAL BOTANY, that branch of economic botany which treats of weeds and cultivated plants. It includes the systematic study of such plants and their methods of reproduction, the laws of improvement and degeneration in plants, etc., also vegetable pathology in its application to cultivated plants.

ĂGRŎSTŎG'RĂPHY̆, see Agrostology.

ĂGRŎSTŎL'ŎGY̆, the part of botany relating to grasses.

ĂG'Y̆NOŬS, without pistils.

ĀI'GRET, any feathery crown or tuft attached to the seed, as the coma of the milkweed or pappus of the thistle; aigrette; egret.

ĂIGRETTE', see Aigret.

AIR-BLADDER, an organ filled with air for the purpose of floating the plant in water, as in the sea-weed Fucus; air-sac; air-cell.

AIR-CHAMBER, see Stomatic Chamber and Air-passage.

AIR-PASSAGE, an extended opening between the cells containing air, as in the stems of many water-plants; lacuna; air-chamber.

AIR-PLANT, a plant growing in the air detached from the soil, as certain orchids; ærophyte. Generally applied only to epiphytic flowering plants.

AIR-FORE, see Stoma.

ĂKENE' see Achenium.

ĀLĂ (pl. Ālæ), see Wing.

ĂLĂBĂS'TRŬM, an old term for flower-bud.

Ă'LAR, (1) borne in the forks of a stem; (2) relating to or having wings. See Alate.

Ă'LĀTE, see Winged.

Ā'LĀTE-PĬN'NĀTE, pinnate with a winged petiole.

ĂLBĔS'ÇENT, whitish; candicant.

ĂL'BĬCĂNT, see Albescent.

ĂL'BĬNĬSM, see Chlorosis.

ĂLBĪ'NO, a plant or variety nearly destitute of chlorophyll, or of which the fruit is abnormally white or colorless.

ĂLBŪ'MEN, nutritive material in many seeds surrounding or adjoining the embryo, endosperm or perisperm.

ĂLBŪ'MEN CRȲSTĂLS, see CRYSTALLOIDS.

ĂLBŪ'MĬNOID, an organic substance containing nitrogen in its composition, as protoplasm; proteid.

ĂLBŪ'MĬNOŪS, furnished with albumen.

ĂLBŬR'NOŪS, having, consisting of, or pertaining to, alburnum.

ĂLBŬR'NŬM, sap-wood; a somewhat distinct, usually lighter colored, outer zone of wood in many exogenous trees and shrubs.

ĂLĔCTŎ'RĬOĪD, filiform, like the thallus of the genus Alectoria in lichens.

ĂLEŪ'RŌNE, proteine-grains which replace starch in the cotyledons or albumen of certain oily seeds.

ĂLGŎL'ŎGȲ, the part of botany relating to algæ.

ĂLĬF'EROŪS, having wings.

ĂL'ĬFŌRM, wing-shaped.

ĂLĬG'EROŪS, see ALIFEROUS.

ĂLLĂN'TŌID, sausage-shaped; botuliform; narrowly oblong; cylindrical with somewhat hemispherical terminations.

ĂLLĂSSŌTŎN'ĬC MOVEMENTS, a term applied by Vries to the movements of mature organs, as the sleep of plants; movements of variation. Compare AUXOTONIC MOVEMENTS.

ĂLLĬĀ'ÇEOŪS, having the odor or other qualities of the genus Allium, which includes the onion and garlic.

ĂLLĪ'ĂNÇE, see COHORT.

ĂLLŎG'ĂMOŪS, habitually cross-fertilized.

ĂLLŎG'ĂMȲ, cross-fertilization. Compare AUTOGAMY.

ĂLLŎT'RŎPĬSM, appearance under an unusual form.

ĂLPĔS'TRĪNE, growing on high mountains below the timber-line, or on the tops of inferior mountains; mountainous.

ALPHĬTŎMŎR'PHOŪS, resembling barley meal. said of certain fungi (Rare.)

ĂL'PĪNE, growing on mountains above the timber-line.

ĂLTĔR'NĀTE, applied to leaves, indicates one at a node; applied to parts of the flower, indicates that the members of one whorl are placed opposite the intervals between the members of the next whorl.

ĂLTĔR'NĀTELȲ-PĬN'NĀTE, pinnate with the pinnæ or leaflets alternating on opposite sides of the rachis.

ĂLTERNĀ'TION OF GENERATIONS, the growth of reproductive bodies into structures differing from that on which they were produced, to return after one or more generations to the original stage or form. Thus, the spore of ferns produces a small prothallus, and upon this sexual bodies are borne which after fertilization reproduce the original fern-plant. See METAGENESIS.

ĂLŪTĀ'ÇEOŪS, of a pale brown color; resembling soft tanned skin in color or texture. (Rare.)

ĂL'VEŎLĀTE, deeply pitted so as to resemble honey-comb, like the receptacle of many Compositæ; faveolate; favose.

ĂMBĬG'ENŪS, having the outer surface of the perianth resembling a calyx and the inner surface a corolla.

ĂMBĬG'ŬOŬS, indistinct or doubtful, so that it cannot well be referred to any definite condition or place in a system of classification, as a bract which has nearly the appearance of an ordinary leaf, or a species which is doubtfully of higher rank than a variety, or one which it is difficult to determine into which of two genera it should be placed.

ĂMBĬP'ĂROŬS, producing two kinds, as a bud which produces directly both flowers and leaves.

ĂM'ENT, a slender spike of naked and usually separated flowers with imbricated scales or bracts; amentum; julus; catkin. Staminate aments are usually deciduous.

ĂMĔN'TĂ, pl., see AMENTUM

ĂMENTĀ'ÇEOŬS, resembling, consisting of, pertaining to, or bearing aments; as an *amentaceous* inflorescence or plant.

ĂMĔNT'ĬFÔRM, amentaceous; juliform.

ĂMĔN'TŬM (pl. **Ămĕn'tȧ**), see AMENT.

ĂM'ĬDŎPLĂST, see LEUCOPLAST.

ĂMMŎPH'ĬLOŬS, growing in sandy places.

ĂM'NĬŎS, the contents of the embryo-sac before the formation of the embryo. Seldom used.

ĂMŒ'BOID, assuming various shapes, like the Amœba

ĂMŎR'PHOŬS, without definite form, structure, or position.

ĂMPHĂN'THĬŬM, see CLINANTHIUM.

ĂM'PHIASTER, a term for the combined nuclear spindle and cytasters in karyokinesis when the latter are present, which is rare in plants. The term is also applied to the combined cytasters only.

ĂMPHĬB'ĬOŬS, growing readily either in water or upon dry land.

ĂMPHĬB'RŸOŬS, growing by additions over the whole surface.

ĂMPHĬCĂR'PĬC, producing two kinds of fruit, either as regards form or period of ripening; amphicarpous. Compare HETEROCARPOUS.

ĂMPHĬCĂR'POŬS, see AMPHICARPIC.

ĂMPHĬG'ĂMOŬS, see AGAMIC.

ĂMPHĬGĂS'TER, used by Bennett and Murray for Amphigastrium.

ĂMPHĬGĂS'TRIĂ (sing. **Amphigastrium**), peculiar scale-like leaves accompanying those of ordinary form as a third row upon the under side of the stem in certain Hepaticæ.

ĂMPHĬG'ĔNOŬS, growing on either surface of a leaf. Said chiefly of certain parasitic fungi.

ĂMPHĬSĂR'CĂ, any indehiscent fruit, hard and dry externally and pulpy within, as a gourd. (Rare.)

ĂMPHĬSPĔR'MOŬS, closely investing the seed so as to have the same form, as the ovary in Gramineæ.

ĂMPHĬT'RŎPAL, see AMPHITROPOUS.

ĂMPHĬT'ROPOŬS, having the funiculus attached to the ovule or seed for half the distance between the chalaza and micropyle; semianatropous; half-anatropous; half-inverted; hemitropous; heterotropous transverse; amphitropal. Applied to the embryo it means curved so that both ends are brought close together.

ĂM'PHŎRĂ, the lower division of a pyxis.

ĂMPLĔC'TĂNT, embracing or clasping, as tendrils, or the sheath of grasses.

ĂMPLĔX'ĬCAUL, partially surrounding or clasping the stem, as the base of many leaves. Compare SHEATHING and PERFOLIATE.

ĂM'PLĬĀTED, enlarged, or moderately dilated.

ĂMPŬL'LĂ, see BLADDER.

ĂMPŬLLĀ'ÇEOŬS, inflated, and swelling out toward the base like a bladder or short flask. Compare LAGENIFORM.

ĂMȲG'DĂLĬNE, pertaining to or resembling the almond.

ĂMȲLĀ'ÇEOŬS, pertaining to, composed of, or resembling starch.

ĂMȲLŎĠĔN'ESĬS, starch-formation.

AMȲLŎGĔN'ĬC, starch-forming. Applied to chlorophyll-granules and similar bodies which originate starch.

ĂM'ȲLOID, a colloid substance having nearly the properties of boiled starch which is found in many sea-weeds and in the seeds of the bean, the almond, etc.

ĂMȲLŎLȲT'ĬC, pertaining to the transformation of starch into other substances, as *amylolytic* diastase.

ĂM'ȲLŬM, starch.

AMYLUM BODIES, see PYRENOIDS.

AMYLUM STAR, see STARCH-STAR.

ĂN'ĂBĬX (pl. Ănăb'içēs), a stem like that of many ferns, etc., which continually dies below and grows above. (Rare.)

ĂNĂBŎL'ĬC, applied by Geddes to the series of ascending metabolic changes in protoplasm by which food is assimilated. Compare KATABOLIC. See ASSIMILATION.

ĂNĂCĂN'THOŬS, without spines.

ĂNĂCRŎĠ'ȲNOŬS, applied in Jungermannieæ by Leitgeb to forms in which the archegonia do not arise upon or near the apex of the shoot, which therefore usually continues to grow after their formation. Compare ACROGYNOUS.

ĂNĀËRŎ'BĬĂ (sing. **Anaërobium**), organisms unable to live or thrive in the presence of free oxygen. Applied to certain bacteria. Compare AEROBIA.

ĂNĀËRŎB'ĬC, see ANAËROBIOTIC.

ĂNĀËRŎBĪŎT'ĬC, being unable to live in contact with air or free oxygen, as some bacteria, or capable of living in an atmosphere destitute of oxygen; anaërobious; anaërobic.

ĂNĀËRŎB'ĬOŬS, see ANAËROBIOTIC.

ĂNĀË'RŎPHȲTE, a plant which does not need a direct supply of air.

ĂN'ĂLŎGUE, an organ or body resembling or having the function of another with which it is compared.

ĂNĂL'ŎĠȲ, resemblance in certain respects, as in general appearance or function. Applied to organs or to classes of plants. Compare AFFINITY, HOMOLOGY, and MORPHOLOGY.

ĂNĂL'ȲSĬS, the systematic examination of a plant preliminary to determining its position in the classification. Compare DETERMINATION.

ĂNĂMÔRPH'ĬSM, see ANAMORPHOSIS.

ĂNĂMÔR'PHŎSĬS, (1) a gradual change of form (generally ascending) traced in a group of plants the members of which have succeeded each other in point of geological time; (2) a similar gradation of form between the members of a group now existing; (3) a remarkable or profound alteration of form resulting immediately from a change in the conditions of growth.

ĂNĂN'DROŬS, without stamens.

ĂNĂN'THĔROŬS, without anthers.

ĂNĂN'THERŬM (obs.), see STAMINODIUM.

ĂNĂN'THOŬS, without flowers.

ĂN'ĂPHAȘEȘ. Strasburger's term for the phenomena of karyokinesis following the metaphases (which see) up to the formation of the resting daughter nuclei.

ĂN'ĂPLĂST, see LEUCOPLAST.

ĂNĂS'TŎMŌSE, to communicate or unite with one another, as the veins of leaves.

ĂNĂSTŎMŌ'SĬS (pl. **Ănăstomō'sēș**), the inosculation or junction of similar parts, often forming a network, as in the veins of leaves.

ĂNĂT'ŎMȲ, VEGETABLE, see VEGETABLE ANATOMY.

ĂNĂT'RŎPOŬS, applied to an ovule or seed which grows so that the funiculus coheres to and forms a raphe along its whole length, bringing the hilum near the foramen and the chalaza at the apparent apex, as in Liliaceæ; anatropal.

ĂNÇĬP'ĬTAL, flattened and two-edged, as the stem of *Panicum anceps;* ancipitous.

ĂNÇĬP'ĬTOŬS, see ANCIPITAL.

ĂNDROCLIN'IUM, see CLINANDRIUM.

ĂNDRŎDĪŒ'ÇIOŬS, having perfect flowers on one set of plants and staminate flowers on another set, but no individuals with pistillate flowers. Compare ANDROMONŒCIOUS and GYNODIŒCIOUS. See POLYGAMOUS.

ĂNDRŒ'ÇĬŬM, the stamens of a flower taken together.

ĂNDRŎGŎNĬD'ĬŬM, see ANDROSPORE.

ĂNDRŎG'ȲNAL, see ANDROGYNOUS.

ĂNDRŎG'ȲNĬȘM, the change from a diœcious to a monœcious condition.

ĂNDRŎG'ȲNOŬS, monœcious with the staminate and pistillate flowers in the same inflorescence. Said mainly of the heads of certain Compositæ. Compare POLYGAMOUS.

ĂNDRŎMŎNŒ'CIOŬS, having staminate and perfect flowers on the same plant, but no pistillate flowers. Compare ANDRODIŒCIOUS. See POLYGAMOUS.

ĂNDRŎPĔT'ĂLOŬS, said of flowers which have become double by the conversion of petals into stamens. (Rare.)

ĂN'DRŎPHŎRE, a column of united filaments, supporting the anthers; stamineal column.

ĂNDRŎPH'ŎRŬM, see ANDROPHORE.

ĂNDRŎSPŎRĂN'GĬŬM, a sporangium containing androspores.

ĂN'DRŎSPŌRE, a kind of asexually produced zoogonidium or swarm-pore in Œdogonieæ which develops into small male plants called "dwarf males;" androgonidium.

ĂN'DROŬS, pertaining to stamens; male.

ĂNEMŎPH'ĬLOŬS, having the pollen or seeds conveyed by the wind.

ĂNFRĂCT'ŪŎSE, bent hither and thither, as the stamens of the melon.

ĂNĠĬŎCĂR'POŬS, having the fruit invested with a calyx, receptacle, or other covering; having spores enclosed by a receptacle of some kind; angiosporous. Especially, having a closed instead of open apothecium in lichens. Compare GYMNOCARPOUS.

ĂNĠĬŎSPĔRM'OŬS, having seeds produced in a closed ovary.

ĂNĠĬŎS'PŎROŬS, having spores or asci produced in a closed receptacle, angiocarpous.

ĂN'GLĔ OF DĒVĬĀ'TION, the angle which a leaf, branch, or root makes with the axis on which it is borne—on the upper side in branches and leaves, on the lower side in root.

ĂNGLE OF DĬVĔR'ĠENÇE, the angle measured by the part of a cycle between succeeding organs in the same spiral or whorl; divergence.

ĂN'GŪLAR, having angles or ridges prismatic. In Pomology, having one side developed more than the other, as in the Newtown Pippin apple. Compare OBLIQUE.

ĂN'GŪLAR DĬVĔR'ĠENÇE, see ANGLE of DIVERGENCE.

ĂN'GŪLĀTE, see ANGULAR.

ĂNGŪLŎDĔN'TĀTE, having angular teeth.

ĂNGŬS'TĀTE, narrow.

ĂNGŬS'TĬFŌ'LĬĀTE, having narrow leaves; angustifolious.

ĂNGŬSTIFŌ'LĬOŬS, see ANGUSTIFOLIATE.

ĂNGŬSTĬSĔP'TĀTE, having a narrow septum or partition, as the pod of shepherd's-purse.

ĂNĪSŎĠ'ŸNOŬS, having fewer carpels than sepals.

ĂNĪSŎM'EROŬS, not having the same number of floral organs in the different whorls. Compare ISOMEROUS and UNSYMMETRICAL.

ĂNĪSŎPĔT'ĂLOŬS, having petals of unequal size. (Rare.)

ĂNĪSŎPHŸL'LOŬS, unequal-leaved, as when the two leaves of a pair are of unequal size.

ĂNĪSŎP'TEROŬS, having wings of unequal size.

ĂNĪSŎSTĂM'ĔNOŬS, see ANISOSTEMONOUS.

ĂNĪSŎSTĔM'ŎNOŬS, having the stamens differing in number from the petals (or parts of the perianth when the sepals are petaloid). Compare ISOSTEMONOUS.

ĂNĪSŎTRŎP'ĬC, having the different parts or organs endowed with different kinds of irritability, or in different degrees, as stems which seek the light and roots which shun it.

ĂNNŎT'ĬNŬS, the ring upon a stem which marks the close of a season's growth. It consists of the scars left by the fallen scales of the bud from which the next year's growth proceeded.

ĂNNŎT'ĬNOŬS, having distinct yearly growths.

ĂN'NŪAL, a plant which usually lives but one year or season.

ANNUAL RING, the ring or cylinder of wood produced by exogenous plants during the year. In some cases the annual ring consists of more than one ring of growth, hence the term "growth-ring" is now often used.

ĂN'NŪLAR, in the form of a circle or zone.

ANNULAR DUCTS, see VESSELS.

ANNULAR VĔS'SELS, vessels marked with thickenings in the form of a ring.

ĂN'NŪLĀTE, marked with rings or circular transverse lines; annulated.

ĂN'NŬLĀTED, see ANNULATE.

ĂNNŪLĀ'TION, a ring or belt.

ĂN'NŪLŌSE, furnished with or composed of rings.

ĂN'NŪLŬS (pl. **Ăn'nūlī**), any body in the form of a ring; the ring or band of thick-walled cells in the sporangium of most ferns which by contraction bursts the sporangium and liberates the spores; the ring of cells upon the inner side of the base of the peristome at the orifice of the theca in many species of mosses which is thrown off to detach the operculum; the ring on the stem of mushrooms; the abortive foliar sheath at the base of the spike in Equisetum.

ĂN'ŎDĂL, see ANODIC.

ĂNŎD'ĬC, applied to the edge of a leaf which a leaf-spiral leaves in its upward course. In a right-handed spiral it would be the right edge of the leaf, and *vice versa*. Compare CATHODIC.

ĂNŎM'ĂLOŬS, unusual in appearance; contrary to rule; abnormal; irregular.

ĂNŎM'ĂLȲ, any deviation from the essential or usual character.

ĂN'SŪLĀTE, coiled at the apex and the whole coil bent over so as to make a loop projecting above the coil, as the growing extremity of the vine of the musk-melon.

ĂN'TECHĀMBER, the upper (outer) angle or space between the guard-cells of a stoma. Compare VESTIBULE.

ĂNTĔN'NÆ (sing. **Ăntĕn'nȧ**), two slender horn-like prolongations of the rostellum in the genus Catasetum in orchids which, upon being touched by an insect or other object, cause the liberation of the retinaculum at their base and the ejection of the pollinium.

ĂNTEPŎSĬ'TION, see SUPERPOSITION.

ĂNTĒ'RĬO-POSTĒ'RIOR PLANE, see MEDIAN PLANE.

ĂNTĒ'RĬŎR, the side of a flower, leaf, or other organ away from the main stem or axis; inferior; exterior; in front. Compare DORSAL.

ĂNTHĔ'LȦ, a paniculate cyme with the lateral axes overtopping the central, as in many species of Juncus and Luzula.

ĂN'THEMȲ, a flower-cluster of any kind. (Obs.)

ĂN'THER, the pollen-bearing part of a stamen.

ĂN'THERĬD'ĬŬM (pl. **Ăntherĭd'iȧ**), the male organ in the higher cryptogams within which antherozoids are produced.

ĂNTHERĬF'EROŬS, anther-bearing.

AN'THER-LOBES, the cells or sacs of an anther, usually two in number, containing the pollen.

ĂN'THERŌID, anther-like.

ĂNTHERŎSPŎRĂN'ĠĬŬM (pl. **Ăntherŏspŏrăn'gĭȧ**), see MICROSPORANGIUM.

ĂN'THERŎZŌID, one of the motile fertilizing bodies produced in an antheridium; spermatozoid.

ĂNTHĒ'SĬS, the time or act of flowering; florescence.

ĂNTHŎCĂR'POŬS FRUIT, generally used for Collective fruit, but more properly for Accessory fruit.

ĂNTHŎCLĬN'ĬŬM, see CLINANTHIUM,

ĂNTHŌ'DĬŬM, the capitulum or head of flowers in Compositæ (popularly called a flower). Compare CLINANTHIUM.

ĂN'THŌID, resembling a flower.

ĂN THŎLĪTE, a fossil flower.

ĂNTHŎL'ȲSĬS, a retrograde metamorphosis of a flower by which normally united or contiguous parts are separated; solution of a flower. The parts of the altered flower may or may not be changed in character.

ĂN'THŎPHŌRE, a stalk or internode sometimes developed between the calyx and corolla, as in Silene.

ĂNTHŎTĂX'ĬS, the arrangement of flowers in an inflorescence; anthotaxy.

ĂN'THŌTĂXȲ, see ANTHOTAXIS.

ĂNTĬCLĪ'NĂL, said of cell-walls or any lines when perpendicular to the outer surface. Compare PERICLINAL.

ĂNTĪ'COŬS, facing anteriorly.

ĂNTĬD'RŎMĂL, twisted or coiled in an opposite direction to that with which it is compared, as when the leaf-spiral of a branch turns in the opposite direction from that of the stem; antidromous. Compare HOMODROMOUS and HETERODROMOUS.

ĂNTĬD'RŎMOŬS, see ANTIDROMAL.

ĂNTĬD'RŎMȲ, having the course of a spiral reversed from the usual direction.

ĂNTĬMȲCŎT'ĬC, see FUNGICIDAL.

ĂNTĬPATHĔT'ĬC, said of plants which do not unite readily when grafted.

ĂNTĬPEDUN'CŪLAR, situated opposite to a peduncle.

ĂNTĬPĔT'ĂLOŬS, situated opposite to a petal, i.e., directly between the petal and the axis; antepetalous.

ĂNTĬP'ŎDĂL CELLS, a group of four cells at the lower end of the embryo-sac, one of which, destitute of a wall, is the lower polar nucleus. Compare EGG-APPARATUS.

ĂNTĬSĔP'TĬC, preventing putrefaction.

ĂNTĬT'RŎPAL, see ANTITROPOUS.

ĂNTĬT'RŎPOŬS, having the radicle of the embryo directed away from the hilum, as in orthotropous seeds.

ĂNTĬZȲMŎT'ĬC, preventing or checking fermentation.

ĂNTRŌRSE', directed upward or forward. Compare ANTICOUS.

ĂPĔRĬSPĔR'MĬÇ, see EXALBUMINOUS.

ĂPĔT'ĂLOŬS, destitute of corolla.

Ā'PEX (pl. **Ā'pĕxes** or **Ăp'ĭcēs**), the extremity opposite the point of attachment.

ĂPHĒLĬOTRŎP'ĬC, turning from the sun; negatively heliotropic.

ĂPHȲL'LOŬS, without leaves.

ĂPHȲL'LȲ, the abnormal suppression of leaves.

ĂP'ĬCAL CELL, the generating cell of a growing point.

ĂP'ĬCAL CONE, see PUNCTUM VEGETATIONIS.

ĂPĬC'ŪLĀTE, terminated by an abrupt short acute point.

ĂPĬC'ŪLĀTED, see APICULATE.

ĂPĬC'ŪLŬS (pl. **Ăpĭc'ūlī**), a small acute point.

ĂPLĂN'OSPORE, applied by Wille to non-motile reproductive cells formed by rejuvenescence in Conjugateæ and some other green algæ.

ĂPLĂS'TĬC, not capable of being organized, or converted into animal or vegetable tissue.

ĂPŎCĂR'POŬS, having the carpels separate or easily separable, as in Ranunculus; dialycarpous.

ĂPŎG'ĂMY̆, habitual non-sexual reproduction, especially vegetative reproduction where sexual reproduction usually occurs, as in the budding of a prothallus in ferns. Compare PARTHENOGENESIS and VEGETIVE APOGAMY.

ĂPŎGĒŎTRŎP'ĬC, growing away from the earth, as ordinary stems.

ĂPŎPĔT'ĂLOŬS, having the petals of the corolla unattached to each other; eleutheropetalous; polypetalous.

ĂPŎPHY̆L'LOŬS, sometimes used for aposepalous, especially in flowers having but one perianth whorl.

ĂPŎPH'Y̆SĀTE, having an apophysis.

ĂPŎPH'Y̆SĬS, an enlargement of the seta below the theca in certain mosses; a thickening on the scales of the cones of certain pines; any irregular swelling.

ĂPŎSĔP'ĂLOŬS, having the leaves of the calyx unattached to each other.

ĂPŎS'TĂSĬS, a term applied by Engelmann to the separation of organs by an unusual extension of the internodes; solution. Compare DIALYSIS.

ĂPŎS'TRŎPHĒ, the collection of the protoplasm and chlorophyll grains along the side walls of a cell instead of the outer surface—often caused by cold or excess or deficiency of light. When caused by deficiency of light, as at night, it is called "negative apostrophe"; when caused by too bright a light, "positive apostrophe." Compare EPISTROPHE and SYSTROPHE.

ĂPŎTHĒ'ÇĬŬM (pl. **Apothecia**), the ascocarp in lichens. See DISCOCARP, SHIELD, and PERITHECIUM.

ĂPPĔND'AGE, any superadded or subordinate part; as hairs, prickles, leaves, etc., upon a stem.

ĂPPENDĬC'ŪLĀTE, having an unusual appendage or appendages, as a winged petiole, or spurred corolla.

ĂP'PLĂNĀTE, flattened out or horizontally expanded. Compare EXPLANATE and COMPLANATE.

ĂP'PLĬCĀTĬVE, see CONDUPLICATE.

ĂP'PŌSĬTE, close together; side by side.

ĂPPRĔSSED', lying close, as leaves to a stem; adpressed.

ĂPPRŎX'ĬMĀTE, close together but not united. Compare REMOTE.

ĂP'TEROŬS, having no dilated appendage or wing. Compare ALATE.

ĂQUĂT'ĬC, growing in water or wet soil. See PALUSTRINE.

Ā'QUEOŬS, nearly colorless. See HYALINE.

ĂRĂCH'NOID, covered with long and loosely entangled hairs, the hairs fewer and longer than in Tomentose; cobwebby.

ĂRĀ'NĒŬS, see ARACHNOID.

ÄR'BŎR, see TREE.

ÄRBŌ'REAL, pertaining to trees or forests.

ÄRBŌ'REOŬS, having the nature of a tree; pertaining to trees.

ÄRBŎRĔS'ÇENT, tree-like in size or form.

ÄRBŎRĒ'TŬM, a botanical tree garden.

ÄR'BŬSÇLE, a low shrub having the form of a tree.

ÄRÇĔS'THĬDȦ (obs.), see GALBULUS.

ÄRCHĒBĪŌ'SĬS, the origin of life. See SPONTANEOUS GENERATION.

ÄRCHĒGŌ'NĬŬM (pl. **Ärchegō'niȧ**), the female organ in the higher cryptogams. Compare OOGONIUM. See ANTHERIDIUM.

ÄRCHĒSPŌ'RĬŬM, the cell, group, or layer of cells from which the spore-mother-cells, and tapetum if any, are derived in the higher cryptogams and flowering plants.

ÄRCH'ETȲPE, the original type or condition.

ÄRCH'ĬCȦRP, the beginning of a fructification; a cell or group of cells fertilized by a sexual act, as an ascogonium or carpogonium.

ÄR'CŪĀTE, curved like a bow.

Ā'RĒȦ, a rather large space bounded by cracks. lines, veins, or part differing in color or texture. Compare AREOLA.

ĂRĔNĀ'ÇEOŬS, growing in sandy places; arenareous; arenose; sabuline; sabulose.

ĂRĔNĀ'RĬOŬS, see ARENACEOUS.

ĂR'ENŌSE, see ARENACEOUS.

ĂRĒ'ŎLȦ (pl. Ărē'ŏlæ), diminutive of area, any small space surrounded by a part differing in structure or color, as the spaces bounded by the veins in reticulate leaves, by the cracks in the surface of the thallus of certain lichens, or by the cell-walls in the leaves of mosses; areolation; areole.

ĂRĒ'ŎLĀTE, divided into areolæ.

ĂRĔŎLĀ'TION, (1) see AREOLA; (2) the form and arrangement of the areolæ in mosses, etc.

Ā'RĒŌLE, see AREOLA.

ÄRĠĔN'TEŬS, white with a tinge of gray; silvery.

ÄRĠĬLLĀ'ÇEOŬS, growing in a clayey soil; argillose.

ÄR'ĠĬLLŌSE, see ARGILLACEOUS.

ĂRHĪ'ZAL, without roots.

ĂR'ĬL, a false coat which sometimes surrounds the seed, growing from the funiculus, hilum, or placenta, as the mace of nutmeg. Compare STROPHIOLE.

ĂRĬL'LȦ, see ARIL.

ĂR'ĬLLĀTE, having an aril.

ĂR'ĬLLĀTED, see ARILLATE.

ĂR'ĬLLŌDE, a false covering to a seed resembling an aril.

ĂRĬL'LŬS, see ARIL.

ȦRĬS'TȦ, see AWN.

ȦRĬS'TĀTE, having an awn.

ĂRĬS'TŪLĀTE, having a small awn.

ÄRM, in horticulture, a large branch of a vine trained horizontally.

ÄRMED, having thorns, spines, or prickles.

ĂRMĬL'LÅ, an annulus in the form of a plaited frill suspended from the top of the stipe below the cap in certain mushrooms.

ĂRŌ'MÅ, a pleasant characteristic odor.

ĂRŌMĂT'ĬC, possessing aroma, especially if spicy.

ĂRRĔCT', directed upward from an inclined base, as the pods of milkweed. (Asclepias.)

ĂR'RŌW-HĔAD'ED, see SAGITTATE.

ĂR'RŌW-SHĀPED, see SAGITTATE.

ÄR'THŎNŌĪD, resembling the apothecium of the genus Arthonia in lichens.

ÄR'THRŌSPŌRE, used mainly in bacteria for a spore formed by segmentation, as opposed to Endospore.

ÄRTHRŌSTĒRIG'MÅ (pl. **Ärthrōsterĭg'mătå**), a jointed sterigma in many lichens, composed of a row of cells from each of which spores are abstricted.

ARTĬC'ŪLĀTE, jointed; articulated.

ARTĬC'ŪLĀTED, see ARTICULATE.

ÄRTĬCŪLĀ'TION, (1) a node, joint, septum, or separable place; (2) one of the segments so marked off or separated.

ÄRTĬC'ŪLŬS, formerly used both for node and internode.

ÄRTĬFĬ'ÇIĂL SYSTEM, a system of classification based on one or a few features only, and not intended to show true relationship, as that of Linnæus.

ĂRŬNDĬNĀ'ÇEOŬS, reed-like.

ĂSÇĔND'ENT, see ASCENDING.

ĂSÇĔND'ĬNG, rising obliquely, or curving upwards from near the base, as the stems of Stellaria or the branches of Norway spruce; adscendent; ascendent: assurgent; adsurgent. Said of ovules which are attached to the middle portion of the placenta or sides of the ovary and are directed upward. Also, directed upward in any manner in contrast to descending, as the *ascending axis*.

ĂSÇĔND'ĬNG ĂXĬS, the stem.

ĂSÇĔND'ĬNG MĔTĂMÔR'PHOSĬS, see PROGRESSIVE METAMORPHOSIS.

ĂS'ÇĪ, pl., see ASCUS.

ĂSÇĬD'ĬÅ, pl., see ASCIDIUM.

ĂSÇĬD'ĬFÔRM, ascidium-shaped.

ĂSÇĬD'ĬŬM (pl. **Ăsçĭd'ĭå**), a hollow pitcher-shaped leaf, like that of Sarracenia.

ĂSÇĬF'EROŬS, see ASCOPHOROUS.

ĂSÇĬG'ĔROŬS, see ASCOGENOUS.

ĂS'CŌCÄRP, the sporocarp of Ascomycetes. See APOTHECIUM, PERITHECIUM, and CLEISTOCARP.

ĂSCŎG'ENOŬS, producing asci; ascigerous.

ĂS'CŌGŌNE, see ASCOGONIUM.

ĂSCŌGŌ'NĬŬM, the carpogonium, or female organ in Ascomycetes before fertilization.

ĂS'CŌPHŌRE, the ascus-bearing layer of hyphæ lining an ascocarp.

ĂSCŎPH'ŌROŬS, ascus-bearing; asciferous. Compare ASCOGENOUS.

ĂS'CŌSPŌRE, a spore produced in an ascus. Often inappropriately called sporidium or sporule.

ĂS'CŬS (pl **Ăs'çī**), one of the characteristic spore-cases of Ascomycetes borne in an ascocarp. It consists of an enlarged terminal cell containing free spores, most frequently eight in number.

ĂS'CŬS-ĂPPĂRĀ'TŬS, the asci, together with the ascogenous cells. See ENVELOPE-APPARATUS.

ĂSCY̆'PHOŬS, without scyphi.

ĂSĔP'TĬC, not liable to putrefaction.

ĂSĔX'ŪAL, without sex; destitute of stamens and pistils in flowering plants, or other sexual organs in cryptogams.

ĂSĔXŪAL ĠĔNERĀ'TION, the second stage or generation in plants having an alternation of generations. It produces spores asexually, but is itself the result of fertilization. In ferns the prothallus is the sexual, and the leaf-bearing plant the asexual, generation. See SPOROGONIUM.

ĂS'PERĀTE, see SCABROUS.

ĂS'PĔRĠĬL'LĬFÔRM, resembling an aspergillus (a round brush used to sprinkle holy water in Roman Catholic churches), as the stigmas of some grasses. Compare MUSCARIFORM.

ĂSPĔRĬFŌ'LĬĀTE, having leaves rough to the touch.

ĂSPĔRĬFŌ'LĬOŬS, see ASPERFOLIATE.

ĂSPĔR'MOŬS, without seeds.

ĂS'PĔROŬS, see SCABROUS.

ĂSSĬMĬLĀ'TION, the conversion of foreign material into the substance of the plant; constructive metabolism.

ĂSSÛR'ĠENT, see ASCENDING.

ĂS'TER, see MOTHER-STAR, DYASTER, AMPHIASTER, and CYTASTER.

ĂS'TEROĬD, having flowers like the Aster. Compare ACTINIFORM.

ĂS'TĬCHOŬS, not in rows.

ĂS'TŌMOŬS, without aperture, as mosses which have not a deciduous operculum.

ĂSY̆MMĔT'RĬCAL, not symmetrical, which see.

ĂT'ĂVĬSM, resemblance to a distant ancestor; remote heredity.

Ā'TER (in composition, *ătro-*), pure black, as distinguished from Niger.

ĂTHĂL'LĪNE, without thallus.

ĂT'RŌPĂL, see ORTHOTROPOUS.

ĂT'RŌPHY̆, a dwarfed or stunted condition of an organ as compared with others, or with the usual state. Especially applied to parts which seem wasted away from lack of nourishment or other cause. Compare ABORTION and DEGENERATION.

ĂT'RŌPOŬS, see ORTHOTROPOUS.

ĂTRŌPÛRPŪ'REŬS, dark purple.

ĂTTĔN'ŪATE, tapering gradually to a point or narrow extremity.

ĂTTĔN'ŪATED, see ATTENUATE.

AUGMĔNTĀ'TION, increase beyond the normal number, especially the production of additional floral whorls.

AU'LŌPHY̆TE, a plant which lives within another for shelter only, not as a parasite, as some Protococcaceæ.

AURĂNTĬĀ'ÇEOŬS, orange-colored. Darker than Aureus.

AU'REŬS, yellow, with a slight admixture of red; golden.

AU'RĬCLE, any ear-like appendage, as the lobes at the base of the leaves in sorrel.

AU'RĬCLED, see AURICULATE.

AURĬC'ŪLĂ (pl. Aurĭc'ūlæ), see AURICLE.

AURĬC'ŪLĀTE, having auricles, or ear-like lobes or appendages.

AURĬC'ŪLĀTED, see AURICULATE.

AU'RĬFÔRM, having the form of the human ear.

AUSTĔRE', astringent or harsh to the taste.

AUTŌCĂR'POŬS, said of ovaries which are not adherent to the calyx; superior.

AUTŒ'ÇIOŬS, said of a parasitic fungus which inhabits the same host-plant through all its stages of growth. Compare HETERŒCIOUS. Applied in mosses when the male and female "flowers" are in separate involucres upon the same plant.

AUTŌG'ĂMOŬS, self-fertilizing.

AUTŌG'ĂMY̆, close-fertilization; the fertilization of a flower by its own pollen. Compare ALLOGAMY.

AUTŌĠĔN'ĔSĬS, see SPONTANEOUS GENERATION.

AUTŎĠ'ĔNOŬS, self-originating. Applied to diseases which have their origin or cause within the effected organism. Compare INFECTION, ESOTERIC, and EXOTERIC.

AUTŎĠ'ENŬS or **MONŎĠ'ENŬS**, terms proposed in place of Monotypic, to indicate that a genus contains but a single species.

AUTŎN'ŌMOŬS, complete in itself. Applied to forms which continually and directly reproduce themselves, and are not mere stages in the life of a plant.

AUTŌPHY̆LLŎĠ'ENY̆, the growth of one leaf upon another.

AU'TŌPLĂST, see CHLOROPHYLL-BODY.

AUTŌTĔM'NOŬS, capable of spontaneous division, as ordinary growing cells.

AUX'ŌSPŌRE, a large renewal-cell in Diatomaceæ, formed either by rejuvenescence, or developed from a zygospore produced by conjugation.

AUXŌTŎN'IC MOVEMENTS, those made by growing organs, as the twining of stems. Compare ALLASSOTONIC MOVEMENTS.

ĂVĔRSE', turned or facing away from the central axis or other object. Compare ADVERSE.

AWL'-SHĀPED, see SUBULATE.

AWN, a bristle-shaped appendage, like those on the glumes of many grasses; arista; beard.

AWNED, having an awn or beard; aristate.

ĂX'-FÔRM, see DOLABRIFORM.

ĂX'ĬAL, pertaining to an axis; extending in the direction of an axis; forming an axis. Compare AXILE.

ĂXĬF'EROŬS, having an axis; producing stems but no leaves.

ĂX'ĬL, the upper angle between leaf and stem; any angle.

ĂX'ĬLE, occupying or belonging to the axis, as an *axile placenta.*

AXĬL'LA (pl. Ăxil'læ), see AXIL.

ĂX'ĬLLAR, see AXILLARY.

ĂX'ĬLLĀRY̆, pertaining to or occupying an axil.

ĂX'ĬS, the central line of any body; an organ around which others are attached, especially a main stem or root.

AXIS, ASCENDING, see ASCENDING AXIS.

AXIS, DESCENDING, see DESCENDING AXIS.

AXIS OF ĬNFLORĔS'ÇENÇE, the part of the stem or branch along which flowers are borne. See RECEPTACLE and RACHIS.

AX'-SHĀPED, see DOLABRIFORM.

ĂZȲG'ŌSPŌRE, a spore in certain algæ and fungi resembling a zygospore, but produced asexually.

ĂZ'ȲGOŬS, without a fellow or corresponding part, as a leaflet which does not have another leaflet on the opposite side of the rachis.

BĂC'CÀ, see BERRY.

BĂC'CĀTE, berry-like; pulpy.

BĂC'CĀTED, covered with berries, or bodies resembling berries.

BĂCÇĬF'EROŬS, producing berries.

BĂC'ÇĬFÔRM, see BACCATE.

BĂ'ÇĬLLÀR, rod- or club-shaped, like a bacillus; bacilliform.

BĂCK, see DORSUM.

BĂCTĒ'RIŌĬDS, protoplasmic bodies endowed with Brownian movement, and resembling bacteria.

BALD, destitute of the usual covering, as of hair, foliage, etc.; destitute of beards or awns.

BĂLŬS'TRÀ, sometimes applied to fruits like the pomegranate

BĂND, (1) a space between two ridges on the fruit of Umbelliferæ; (2) a broad stripe, especially if transverse.

BĂN'NER, see VEXILLUM.

BÄRB, a sharp reflexed point on an awn or other process; a hair or other process having such reflexed points, or with a reflexed tip.

BÄR'BĀTE, see BEARDED.

BÄR'BĀTED, see BEARDED.

BÄRBE, see BARB.

BÄR'BELLĀTE (diminutive of Barbate), having minute barbs; barbellulate.

BÄRBĔL'LŪLĀTE, see BARBELLATE.

BÄR'BŪLE, a small barb.

BÄRK, the covering of the stem and roots of exogens, separated from the wood by the cambium.

BÄRRED, crossed by parallel horizontal bands or lines.

BĂR'REN, unproductive; unfruitful; sterile,—said of a plant or organ. The term "barren" is seldom applied except to a plant as a whole, and even then the term "sterile" is more often used.

BĀ'SAL, pertaining to the base.

BĀSE, the part of an organ by which it is attached to its support.

BĀSĬDĬŌĠENĔT'ĬC, borne on a basidium.

BĀSĬD'ĬŌPHŌRE, a sporophore bearing a basidium.

BĀSĬD'ĬŌSPŌRE, a spore borne on a basidium, as those of mushrooms.

BĀSĬD'ĬŬM (pl. **Băsĭd'ĭà**), one of the branched cells on the spore-bearing surface of mushrooms, etc., which bear the spores. Compare STERIGMA.

BĀS'ĬFĬXED, attached by its base, as an innate anther.

BĀSĬF'ŪGĂL, (1) proceeding away from the base; upward; acropetal; centrifugal (centripetal when applied to the order of inflorescence); (2) derived from the base or situated at the base, as *basifugal* growth in the leaves of certain grasses.

BĂSĬĠȲN'ĬUM, see GYNOPHORE.

BĀS'ĬLAR, relating to or situated at the base.

BĀ'SIN, the depression at the

apex of an apple. Compare CAVITY and EYE.

BĀSĬP'ETĂL, proceeding or produced in succession toward the base; downward; centrifugal (when applied to the order of inflorescence only). Compare ACROPETAL.

BĀSĬSCŌP'ĬC, on the side toward the base; facing the base.

BȦSS, see BAST.

BĂST, the liber or inner bark; particularly the long, tapering, thick-walled phloëm-cells characteristic of such bark.

BĂS'TÄRD, spurious; resembling something else. Used improperly for Hybrid.

BĂST-ÇĔLLS, thick-walled elongated spindle-shaped cells in the phloëm portion of fibro-vascular bundles. They give strength and flexibility to the tissues; bast-fibres.

BĂST-FĬBRES, see BAST-CELLS.

BAST-SHĔATH, see PHLOEM-SHEATH.

BAST, SOFT, see SOFT BAST.

BAST-TISSUE, see PHLOEM

BAST-VĔS'SEL, see SIEVE-TUBE.

BĔAK, a process like the beak of a bird terminating the fruit of many leguminous and other plants; rostrum.

BĔAKED, ending in a prolonged narrow rigid tip like a beak; rostrate.

BĔARD, the awns of grasses, singly or collectively; a tuft of hairs, especially if stiff and long; sometimes applied improperly to the lower lip of labiate corollas.

BĔARD'ED, having a beard; beset with hairs, especially if stiff and long.

BĒARD'LETED, having minute beards or awns.

BĔLL'-SHĀPED, see CAMPANULATE.

BĔL'LȲĬNG, swelling out on one side, as the tube of the corolla in many Labiateæ.

BĔR'RĬED, see BACCATED.

BĔR'RȲ, an indehiscent fruit pulpy or fleshy throughout, as the grape, currant, and tomato.

BĪĂCŬ'MĬNĀTE, two-pointed.

BĪĂN'GŪLĀTE, having two angles or corners.

BĪARTĬC'ŪLĀTED, having two joints.

BĪẠURĬC'ŪLĀTE, two-eared.

BĬBRĂC'TEĀTE, with two bracts.

BĬBRĂC'TEŌLĀTE, having two bractlets.

BĪCĂL'CĂRĀTE, having two spurs.

BĪCĂL'LŌSE, having two small hard spots or protuberances.

BĪCĂP'SŪLAR, having two capsules. Sometimes applied to capsules which are divided before dehiscence into two parts.

BĪCÄR'PELLÄRȲ, see DICARPELLARY.

BĪCĂR'ĬNĀTE, two-keeled, as the palet of grasses.

BĪÇĔPH'ĂLOŬS, having two heads.

BĪÇĬP'ĬTĂL, divided into two parts.

BĪCĬP'ĬTOŬS, see BICIPITAL.

BĪCOLLĂT'ERAL BUNDLE, a fibro-vascular bundle in which a strand of xylem lies between two strands of phloem.

BĪ'COLŎR, of two colors.

BĪ'COLŌRED, see BICOLOR.

BĪCŎN'JŪGĀTE, see BIGEMINATE.

BĪCÔR'NŪTE, with two horn-like processes.

BĪCRĒ'NĀTE, with two crenatures or rounded teeth. Compare DOUBLY-CRENATE.

BĪCRU'RAL, having two legs or narrow elongations, as the lip of the "man-orchis."

BĪCŬS'PĬD, having two sharp stiff points or cusps; bicuspidate.

BĪCŬS'PĬDĀTE, see BICUSPID.

BĬDĔN'TĀTE, having two teeth. Compare DOUBLY-DENTATE.

BĬDĬG'ĬTĀTE, having two finger-like divisions.

BĬD'ŪOŬS, lasting two days only.

BĪĔN'NĬAL (adj.), living two years, or requiring two seasons to come to maturity.

BĪĔN'NĬAL, (n.), a plant which usually matures its fruit the second year and then dies.

BĬFĀ'ÇIĂL, applied to leaves which have a distinct upper and lower surface differing in texture. Compare CENTRIC.

BĪFĀ'RĬOŬS, pointing in two directions; two-ranked; distichous.

BĪ'FĔR, a plant which ripens fruit twice a year.

BĬF'EROŬS, fruiting twice a year.

BĪ'FĬD, divided about to the middle in two parts; two-cleft.

BĬF'ĬDĀTE, see BIFID.

BĪFĬS'TŪLAR, containing two tubular openings.

BĬFLŌ'RĀTE, see BIFLOROUS.

BĬFLŌ'ROŬS, two-flowered.

BĬFŌ'LĬĀTE, having two leaves or leaflets.

BĬFŌ'LĬŌLĀTE, having two leaflets.

BĬFŎLLĬC'ŪLAR, with a double follicle.

BĬFŌ'RĀTE, having two perforations, as the anthers of Rhododendron.

BĬF'ŌRĬNE, a peculiar cell containing raphides found in arums and certain other plants—so called because when placed in water they become turgid and discharge their contents, often from both ends.

BĬF'ŌROŬS, see BIFORATE.

BĪ'FRŎNS, having two faces or aspects; growing on both surfaces of a leaf (amphigenous).

BĬFÛR'CĀTE, forked; divided into two branches.

BĬFÛR'CĀTED, see BIFURCATE.

BĬFÛRCĀ'TION, division into two branches.

BĬFÛR'COŬS, see BIFURCATE.

BĬĠĔM'ĬNĀTE, twice paired, as a decompound leaf with two pairs of leaflets, i.e., having a forked petiole with a pair of leaflets at the end of each division; biconjugate.

BĪ'ĠĔNĔR, see GENUS-HYBRID.

BĪGLĂND'ŪLAR, having two glands or gland-like bodies.

BĬJŪ'GĀTE, having two pairs, as a leaf with two pairs of leaflets.

BĬJŪ'GOŬS, see BIJUGATE.

BĬLĀ'BĬĀTE, see LABIATE.

BĬLĂM'ĔLLÄR, see BILAMELLATE.

BĬLĂM'ELLĀTE, of two plates or lamellæ.

BĬLĂM'ELLĀTED, see BILAMELLATE.

BĬLĂT'ERAL, two-sided.

BĪLŌ'BĀTE, two-lobed.

BĪLŌ'BĀTED, see BILOBATE.

BĪ'LŌBED, see BILOBATE.

BĪLŎC'ŪLÄR, two-celled—applied to ovaries, anthers, etc.

BĪMĂC'ŪLĀTE, having two spots.

BĪMĂC'ŪLĀTED, see BIMACULATE.

BĪ'NĀRȲ, in twos; double.

BĪ'NĀTE, in twos or pairs; conjugate—said of two bodies of the same nature springing from the same point. In speaking of pinnate leaves the term *conjugate* is generally used.

BĬNE, a twining or climbing stem. Rare except in composition, as *wood-bine.*

BĪNẼRV'ĀTE, having two nerves or veins, or two which are especially prominent.

BĬNŌ'DĂL, containing two nodes only.

BĪNŌ'MĬAL, of two names, as the generic and specific names which compose a "botanical name."

BĪ'NOŬS, see BINATE.

BĪNŪ'CLĒĂR, see BINUCLEATE.

BĪNŪ'CLĒĀTE, have two nuclei or central points; binuclear.

BĪNŪ'CLĒŌLĀTE, with two nucleoli.

BĪŎÇ'ELLĀTE, with two eye-like spots.

BĪŌĠĔN'ESĬS, (1) the origin of life; (2) the production of living beings from other living beings in any manner—the converse of spontaneous generation (abiogenesis).

BĪŎĠ'ENOŬS, growing on living plants, either parasitic or not.

BĪŎĠ'ENỸ, the genesis or evolution of living forms, or the science which treats of it, including Ontogeny and Phyllogeny.

BĪŎL'ŌĠỸ, zoology and botany.

BĪŌLỸT'ĬC, injurious or destructive to life.

BĪ'ŎN, an individual morphologically independent.

BĪŎPH'ĂGOŬS, feeding on living organisms. Applied chiefly to insectivorous plants.

BĪ'ŌPLĂŞM, any living fluid; the same as protoplasm.

BĬPĂL'ĔŌLĀTE, with two lodicules; bilodiculate.

BĬPĂL'MĀTE, said of leaves which are palmate upon secondary palmate petioles.

BĬP'ĀROŬS, bearing two objects, as a leaf with two leaflets.

BĬPÄR'TED, see BIPARTITE.

BĬPÄR'TĬBLE, divisible into two parts.

BĬPÄR'TĬLE, see BIPARTIBLE.

BĬPÄR'TĪTE, two-parted; divided into two parts to the base or nearly so.

BĬPÄRTĬ'TION, the act of dividing into two equal parts. Compare BISECTION.

BĬPĔC'TĬNĀTE, toothed like a comb on two sides.

BĬPEL'TĀTE, having two shield-shaped parts.

BĪPĔRĔN'NĬĂL, said of a part which lives two years, but reproduces itself indefinitely, as the tubers of the potato. (Rare.)

BĪPĔT'ĂLOŬS, having two petals.

BĬPĬN'NĀTE, said of a pinnate leaf with secondary petioles, each bearing more than one leaflet, as the honey-locust; doubly pinnate; twice pinnate.

BĬPĬN'NĀTED, see BIPINNATE.

BĬPĬNNĂT'ĬFĬD, having the divisions of a pinnatifid leaf pinnatifid; twice pinnately cleft.

BĪPĬNNĂTĬPÄR'TĪTE, having the divisions of a pinnatipartite leaf pinnatipartite; twice pinnately parted. It differs from Bipinnatifid in having the divisions extend to near the midrib.

BĪPĬNNĂT'ĬSECT, having the divisions of a pinnatisect leaf pinnatisect; twice pinnately divided. The divisions extend to the midrib, but the segments are sessile.

BĬPLĪ'CĀTE, twice or doubly folded.

BĪPŌ'RŌSE, having two small openings or pores.

BĪPŬNC'TĀTE, having two small spots.

BĪRĀ'DĬĀTE, having two rays.

BĪRĪ'MŌSE, opening by two slits, as most anthers; having two clefts, slits, or narrow openings.

BĬSĂC'CĀTE, having two sacs or pouches.

BĪSCŪ'TĀTE, resembling two round bucklers placed side by side.

BĬSĔCT', to divide into two equal parts.

BĬSĔC'TION, the act of cutting or dividing into two equal parts; bipartition.

BĬSĔP'TĀTE, having two partitions or septa.

BĬSĒ'RĬĂL, in two rows or series.

BĬSĔR'RĀTE, having the teeth or serratures serrate; doubly serrate. The latter term is preferable, and *Biserrate,* though in more common use in this sense, should be applied in analogy with Bidentate to leaves or margins bearing two serratures.

BĬSĒ'TŌSE, having two bristles; bisetous.

BĪSĒ'TOŬS, see BISETOSE.

BĪSĔX'ŪAL, containing both sexes, as a flower with both stamens and pistils; hermaphrodite; monoclinous; synœcious.

BĬSPĪ'NŌSE, having two spines.

BĪ'SPŌRE, a two-spored tetraspore.

BĪSTĬP'ŪLED, having two stipules.

BĪSŬL'CĀTE, having two longitudinal grooves or furrows.

BĪTĔR'NĀTE, divided into three parts, each of which is divided into three.

BĬT'TEN, see EROSE.

BĪ'VĂLVE (adj.), having two valves, as some capsules; bivalvular.

BĪ'VĂLVE (n.), a capsule of two valves.

BĪVĂLV'ŪLĂR, see BIVALVE.

BĪVĂS'CŪLĂR, having two vessels.

BĪVĬT'TĀTE, having two vittæ.

BLĂD'DER, (1) an inflated membranons pericarp; (2) a membranous air-sac in some water-plants which enables them to float. See AIR-BLADDER.

BLĂD'DERȲ, thin and inflated like a bladder, as the calyx of *Silene inflata.*

BLĀDE, the expanded portion of a leaf; lamina. The term *blade* is more commonly applied in grasses and *lamina* in other plants.

BLĂNCHED, whitened by absence of light; etiolated. Compare CHLOROSIS and ALBINISM.

BLĂND, fair; beautiful.

BLĂSTĒ'MȦ, the embryo aside from the cotyledons; also used for any point of growth or budding part. (Rare.)

BLĂSTĒ'MAL, rudimentary; nascent. (Rare.)

BLĂSTŌCĂR'POŬS, applied to a fruit when the seed germinates within the pericarp, as sometimes occurs in the mangrove. (Rare.)

BLĂSTŌCŎL'LȦ, the gummy substance on many buds, as on the horse-chestnut.

BLĂSTŎĠĔN'ESĬS, reproduction by buds; gemmation. (Obs.)

BLĂS'TŬS (obs.), see BUD and PLUMULE.

BLĬND, applied to a malformation, chiefly in certain cultivated plants, as cabbage and cauliflower, in which the stem terminates without producing a head or inflorescence. A "blind bud" is one which fails to develop. To "go blind" is to fail to produce flower-buds where expected.

BLĬS'TERED, see BULLATE.

BLOOM, a coating on the surface of fruits, leaves, etc., often grayish or bluish in color, consisting of minute, waxy particles in the form of filaments, granules, or layers. See GLAUCOUS.

BLŌSSOM-BUD, see FLOWER-BUD.

BLŌTCHED, having distinct irregular spots of color. Compare CLOUDED.

BLŬNT, obtuse.

BŌAT'-SHĀPED, see NAVICULAR.

BŎB, a popular name for the inflorescence (thyrse) of sumach.

BŌLE, the body of a tree.

BŌLL, a globular pericarp, as that of cotton.

BŌLT (Hort.), to run prematurely to seed, as carrots when they seed the first year.

BŌN'Ȳ, hard, brittle, and close in texture, as the stone of the peach.

BOOT, a popular name for the sheath of grains and other grasses.

BÔR'DĒR, the expanded portion of a gamopetalous corolla, consisting of the united limbs. See also BORDERED PIT.

BŌR'DĒRED, having the margin different from the remainder in form, color, or texture.

BŌR'DĔRED PIT, a thin spot or opening in a cell-wall covered on each side by a thickened convex body having a central perforation; areolated dot; discoid marking. Bordered pore of Gregory. These markings are characteristic of the wood-cells of Coniferæ. The *border* is the more or less dilated central portion of the pit or passage between the cells.

BÔR'DĒRED PŌRE, see BORDERED PIT.

BŌSS, a rounded protuberance.

BŌSSED, having a boss. Compare UMBONATE.

BŌS'TRȲCHOID, ÇȲME see HELIÇOID CYME.

BŎS'TRȲCHŌID DĬCHŌT'ŌMȲ, see HELICHOID DICHOTOMY.

BŎS'TRȲX, see HELICOID CYME.

BŌTĂN'ĬCĂL ĠĒOG'RĂPHȲ, see GEOGRAPHICAL BOTANY.

BŌTĂN'ĬCĂL NAME, the generic name followed by the specific name.

BŌTĂN'ĬCĂL NŌMENCLĀT'ŪRE, an account of the names of plants, and of the laws for their application.

BŌTĂN'ĬCĂL TĔRMĬNŎL'ŎĠȲ, an account of the special words used in describing plants.

BŌTĂN'ĬC GÄR'DEN, a collection of growing plants systematically arranged for the purpose of study.

BŌT'ĂNĬZE, to seek for growing plants for the purpose of botanical investigation.

BŌT'ĂNȲ, the science of plants; phytology. See STRUCTURAL, PHYSIOLOGICAL, and SYSTEMATIC BOTANY, BOTANICAL NO-

MENCLATURE, and BOTANICAL TERMINOLOGY.

BŌTHRĘN'CHȲMȦ (obs.), see PITTED TISSUE.

BŎT'RŬS, an old term for a cluster like that of the grape.

BŌT'RŸOID, having the form of a bunch of grapes; botryoidal; botryose.

BŎT'RŸŌSE, see BOTRYOID.

BŌTRŸOI'DĂL, see BOTRYOID.

BŌT'TLE-SHĀPED, see LAGENIFORM.

BŎTŪ'LI-FŌRM, see ALLANTOID.

BRĂCH'ĬĀTE, applied to branches which are decussate and spreading, as in the ash, and sometimes improperly to those which are long and widely spreading, but not decussate.

BRĂCHŸP'ŌDOŬS, short-stalked.

BRĂCT, a leaf of an inflorescence, generally subtending a flower or flower-branch. Bracts are intermediate in position between foliage leaves and floral leaves, and are usually much smaller than the ordinary leaves, and in other respects more or less modified.

BRĂC'TĒȦ (pl. **Brăc'tēæ**), see BRACT.

BRĂC'TĒȦL, of the nature of a bract.

BRĂC'TĒĀTE, furnished with bracts; subtended by a bract or bracts; bracted.

BRĂCTĒ'ŌLȦ (pl. **Bracteolæ**), see BRACTLET.

BRĂC'TĒŌLĀTE, furnished with bractlets.

BRĂC'TĒŌLE, see BRACTLET.

BRĂC'TĒŌSE, having many or conspicuous bracts.

BRĂCT'LĔT, a small bract, or bract upon a pedicel or secondary branch of an inflorescence.

BRĂCT'LĔSS, without bracts.

BRÃIRD, to germinate. (Scotch.)

BRÀNCH (n.), a division of a stem or other elongated organ; secondary axis.

BRÀNCH (v. i.), to produce branches; to ramify.

BRÀNCH'LĔT, a little branch or subdivision of a branch; twig.

BREĀK (v. i.) (Hort.): (1) to depart widely from the type and suddenly produce a new variety (nearly the same as *sport*); (2) to "bolt" or run prematurely to seed, as a biennial the first year; (3) to put forth new buds or leaves.

BREAST-WOOD (Hort.), branches which project outward from a wall or espalier.

BRĔATH'ĬNG-PŌRE, see STOMA.

BREED, see RACE.

BRĪ'DLES, strings of protoplasm which often connect the nucleus with the layer of protoplasm next the cell wall.

BRĬS'TLE, a stiff, short hair or hair-like body.

BRĬS'TLE-POINT'ED, ending in a bristle; terminating gradually in a fine sharp point, as the leaves of many mosses.

BRĬS'TLŸ, beset with bristles.

BROOD'-BUD, a deciduous leaf-bud capable of growing into a new plant, such as the bulbils of the tiger-lily and the deciduous buds of certain lycopodiums; also applied to the soredium of lichens.

BROOD'-CELL, any reproductive unicellular body produced asexually which separates from the parent plant, as the gonidia of fungi.

BROOD'-GĔM'MȦ, see GEMMA.

BROWN'ĬAN MOVEMENT, a trembling movement common to all minute particles suspended in a liquid. Called also Brunonian movement, Pedesis, Non-vital motion, and (improperly) Molecular movement. The cause is not known.

BRŪN'NEŬS, deep brown, a mixture of dark gray and red.

BRŬSH'-SHĀPED, see ASPERGILLIFORM.

BRȲŌL'ŌĠȲ, the division of botany relating to mosses; muscology.

BŬCK'LER-SHĀPED, see SCUTATE.

BŬD, an undeveloped stem or branch, or its extremity, bearing rudimentary leaves which are specially modified for its protection. See LEAF-BUD, FLOWER-BUD, and GEMMA.

BŬD'DING, putting forth buds or gemmæ. Also applied to Pullulation, which see.

BŬD'LĔT, a little bud attached to a larger one.

BŬD-RṲ'DĬMĔNT, the special cells which originate the leaf-bearing axis in the pro-embryo of Characeæ.

BŬD'-SCĀLE, one of the modified leaves of a bud; perule.

BŬD'-SPŌRT, see BUD-VARIATION.

BŬD-VĀRĬĀ'TION, the development of a bud in a manner unusual to the species or variety, and different from the other buds upon the plant. It may be the production of a new variety or a reversion to an earlier form. Bud-variations, or "bud-sports" as they are often called, may usually be propagated by division, but their characters are seldom retained when grown from seed. Compare SEED-VARIATION.

BŬD-VĀRĪ'ETȲ, a variety which originated by bud-variation.

BŪGLE-SHĀPED, having the shape of a bugle bead;—a form varying from oblong to obovoid. Used in describing the fruit of certain cranberries.

BŬLB, a bud with thickened scales containing nutriment for its development, thus differing from ordinary buds, the scales of which are for protection only. Bulbs are usually subterranean.

BŬLBĀ'ÇEOŬS, having bulbs; bulbous.

BŬLBED, in the form of a bulb; bulbaceous.

BŬL'BĔL, see BULBLET.

BŬLBĬF'EROŬS, bulb-bearing; bulbous.

BŬL'BĬL, see BULBLET.

BŬLB'LĔT, a little or secondary bulb, especially one above ground, as in some lilies and ferns; bulbel; bulbil; bulbule. See CLOVE.

BŬL'BŌSE, bulb-like in shape or structure.

BŬL'BŌ-TŪBER, see CORM.

BŬL'BOŬS, producing bulbs, growing from bulbs, or bulbose.

BŬLB-SCĀLE, one of the thickened scale-like leaves of a bulb.

BŬLB, TŪ'NĬCĀTĔD, see TUNICATED BULB.

BŬL'BŬLE, see BULBLET.

BŬL'BŬS, the swollen base of the stipe in mushrooms.

BŬL'LĀTE, having the surface blistered or puckered, as the leaves of the Savoy cabbage, the spaces between the veins of which are concave on one side and convex on the other.

BŬL'LĬFŌRM CĔLLS, see HYGROSCOPIC CELLS.

BŬNCH, an indefinite cluster or tuft.

BŬN'DLE-SHĒATH, a layer of closely united thin-walled parenchyma partly or wholly surrounding each fibro-vascular bundle or continuous around the fibro-vascular cylinder; phloem-sheath.

BÛR, a seed or head bearing hooked or barbed appendages which serve for its attachment to various animals, thus securing its dissemination.

BŪRSĬC'ŪLĀTE, pouch-like or furnished with pouch-like appendages.

BÛRSĬC'ŪLȦ (pl. **Bursic'ulæ**), an old term for the stigmatic chamber in orchids.

BÛRSĬC'ŪLĀTE, having a bursicula or small pouch.

BUSH, a shrub, especially if dense and low. *Shrub* implies the habitual form or limit of growth, but a *bush* may grow into a tree.

BŬT'TERFLȲ-SHĀPED, see PAPILIONACEOUS.

BUTTERY, applied to fruits, especially pears, the flesh of which is soft and yields readily to the pressure of the teeth, as in the White Doyenné and Seckel pears. The texture is in some respects intermediate between Breaking and Melting, which see.

BŬT'TON (v. i.) (Hort.), to form small heads prematurely, as cauliflower.

BĪÇȲC'LĬC, having two cycles or whorls.

BȲSSĀ'ÇEOŬS, resembling or consisting of fine filaments like cobweb, as the mycelium of mushrooms.

BȲS'SOID, resembling bissus.

BȲS'SŬS, an old name for the filamentous mycelium of certain fungi.

CĂDŬ'COŬS, falling early, as the calyx of the poppy.

CÆRU̧LĒS'ÇENT, see CŒRULESCENT.

CÆRU̧'LEŬS, see CŒRULEUS.

CÆ'SIOŬS, pale bluish-gray; lavender-colored.

CÆSPĬT'ĔLLŌSE, diminutive of Cæspitose.

CÆS'PĬTŌSE, see CESPITOSE.

CĂLȦTHĬD'ĬŬM (obs.), see CAPITULUM. Formerly applied to the head of flowers in Compositæ, or sometimes to the involucre only.

CĂLĂTH'ĬFÔRM, bowl-shaped, with the margin more or less flaring, like a fruit-dish or flower-basket.

CĂL'CĂR, see SPUR.

CĂL'CĂRĀTE, having a spur, as the flower of larkspur; spur-shaped.

CĂLCĀ'RĔOŬS, of a dull chalk-white color; growing in chalky or limestone soils.

CĂL'ÇĒĬFŌRM, see CALCEOLATE.

CĂL'ÇĒŌLĀTE, slipper-shaped, as the lip of Cypripedium; calceiform; soleæform.

CĂL'ÇĬFŌRM, powdery, like chalk or lime.

CĂLÇĬV'ŌROŬS, eating into lime rock, as certain lichens.

CĀLĬC'ŪLĂR, cup-shaped. (Rare.)

CĀLĬC'ŪLĀTE, see CALYCULATE.

CĀ'LĬX, see CALYX.

CĂL'LĪ (pl.), see CALLUS.

CĂL'LŌSE, having hardened spots or protuberances; callused.

CĂLLŎS'ĬTȲ, a hard or thickened spot or protuberance; callus; wart.

CĂL'LŬS (pl. **Calluses** or **Calli**), (1) a hard or thickened spot or protuberance; callosity: (2) the new formation upon an injured surface, as seen at the end of a cutting: (3) a thickened deposit of formative material in the pores of the sieve-plates in certain trees in autumn: (4) a term applied to an extension of the flowering glume in grasses below its point of insertion, and which is grown to the axis or rachilla of the spikelet and separated from the free portion by a more or less distinct furrow. It is frequently covered with hairs or bristles (as in Stipa) which serve for attachment to other objects to secure the distribution of the seed. The Callus is sometimes present in the empty glumes also.—(Scribner.)

CĂL'VOŬS, bald; having a surface on which hairs are usually present destitute of them, as in an achenium without a pappus.

CĂLŸCĂN'THĔMOŬS, having petaloid sepals.

CĂLŸCĂN'THĒMŸ, a partial or entire conversion of sepals into petals.

CĂLŸÇĬFLŌ'RĂL, having the calyx free from the ovary, and the stamens (and therefore the petals also) inserted on the calyx. Compare THALAMIFLORAL and COROLLIFLORAL.

CĂLŸÇ'ĬFŌRM, having the form or position of a calyx.

CĂLŸÇ'ĬNAL, see CALYCINE.

CĂL'ŸÇĬNE, pertaining to or situated on a calyx; calycinal.

CĂL'ŸCLE, a whorl of bracts forming a secondary or accessory calyx outside the true calyx.

CĂL'ŸCLED, having a calycle.

CĂL'ŸCŌĬD, resembling a calyx.

CĀLŸC'ŪLĀTE, having a calyculus.

CĀLŸC'ŪLĀTED, see CALYCULATE.

CĀLŸC'ŪLŬS, a set of involucral bracts resembling a calyx, as in Dianthus.

CĀLŸP'TRÀ, the membranous hood or veil covering the capsule in mosses. It consists of the ruptured archegonium carried up by the growing sporophore. The term is also applied to the root-cap, which see.

CĀLŸP'TRĀTE, having a calyptra or similar covering.

CĀLŸP'TRĬFÔRM, shaped like a calyptra or candle-extinguisher, as the calyx of Eschscholtzia.

CĀLŸP'TRŎĠEN, a special layer of cells in certain plants from which the root-cap is developed. (Jancewski.)

CĀ'LŸX, the outer set of floral leaves between the bracts, if any, and the corolla. When there is but one set external to the stamens, it is usually called calyx or perianth.

CĀ'LŸX TŪBE, a tube of united sepals adherent to the ovary or enclosing the other parts of the flower. As the elevated margin of the receptacle sometimes forms a portion of this tube, the term "receptacular tube" is also applied to it.

CĂM'ÀRÀ (pl. **Căm'àræ**), an indefinite term formerly applied to various fruits having more or less membranous carpels, as the Ranunculus and apple. Also applied to a single carpel of such a fruit.

CĂM'BĬAL, pertaining to cambium.

CĂM'BĬFÔRM, said of narrow prismatic thin-walled cells, like those characteristic of cambium.

CĂM'BĬŬM, a layer of meristem belonging to the fibrovascular system, between the xylem and phloem. In exogens it is permanent, and forms the continuous ring or layer of soft growing tissue between the wood and inner bark, from which the new growth of each is derived. It was formerly considered merely a viscid secretion, often called "descending sap," but it is now known to be composed wholly of young, easily broken cells filled with protoplasm or formative matter.

CĂM'BĬŬM LĀY'ER, see CAMBIUM.

CĂM'BĬŬM RĬNG, see CAMBIUM.

CĂMPĂN'ĬFÔRM, see CAMPANULATE.

CĂMPĂNĬL'ĬFÔRM, see CAMPANULATE.

CĂMPĂN'ŪLĀTE, bell-shaped, as the corolla of the harebell.

CĂMPES'TRAL, growing mainly in uncultivated fields.

CĂMPŪLĬT'RŌPOŬS (obs.), see CAMPYLOTROPOUS.

CĂMPȲLŌSPĔR'MOŬS, having the edges of the seed rolled inward to form a longitudinal groove or furrow, as in sweet-cicely and some other Umbelliferæ.

CĂMPȲLŎT'RŌPAL, see CAMPYLOTROPOUS.

CĂMPȲLŎT'RŌPOŬS, applied to ovules or seeds which are curved upon themselves so that the micropyle is near the chalaza, as in Cruciferæ; campylotropal.

CĂNĂL'-ÇELLS, an axial row of cells in the neck of the archegonium the connecting septa of which disappear, forming a canal filled with mucilage for the passage of the antherozoids.

CĂNÀLĬC'ŪLĀTE, channelled.

CĂNÀLĬC'ŪLŬS (pl. **Cănàlic'ūlī**), a channel.

CĂN'ÇELLĀTE, having open network; lattice-like.

CĂN'DĬCĂNT, see ALBESCENT.

CĂN'DĬDŬS, Latin for pure white.

CĀNE, the stem of reeds and other large grasses. Applied in horticulture to the stems of raspberries and blackberries, and one-year-old stems of grape-vines.

CĂNĔS'ÇENT, hoary (gray or whitish) from a coating of fine hairs; canus; incanus; incanous; incanescent.

CĀ'NŬS, see CANESCENT.

CĂP, see PILEUS.

CĂPĬLLĀ'ÇEOŬS, see CAPILLARY.

CĂPĬL'-LĂMENT (rare), see FILAMENT.

CĂP'ĬLLĂRȲ, long and narrow like a hair; said of a filament or channel.

CĂPĬL'LĬFÔRM, see CAPILLARY.

CĂPĬLLĬT'ĬŬM, sterile filaments, often in the form of network, among the spores of puff-balls and some other fungi.

CĂP'ĬTĀTE, furnished with a globose head; growing in a head.

CĂPĬTĔL'LĀTE, diminutive of capitate; capitular.

CĂPĬT'ŪLAR, see CAPITELLATE.

CĂPĬT'ŪLĬFÔRM, in the form of a small head.

CĂPĬT'ŪLŬM (pl. **Căpĭt'ūlă**), a dense flower-cluster, as in the clover or sunflower; a head of any kind.

CĂP'RĒŎLĀTE, having tendrils; cirrhose.

CĂPRĒ'ŎLŬS (pl. **Căprē'ŏlī**), see TENDRIL.

CĂPRĬFĬCĀ'TION, (1) the fertilization of the fig, by hand or by means of insects. Sometimes extended to the artificial fertilization of other fruits. (2) The process of accelerating the ripening of figs by placing on the cultivated plant branches of the wild fig (caprificus). A hymenopterous insect found on the wild plant enters the fruit to lay its eggs, causing it to ripen earlier. The same result is obtained by the practice of pricking the green figs with a needle dipped in olive-oil; also by the application of a drop of the oil to the eye of the fig. There seems to be some doubt as to whether the result from caprification by means of the wild fig is due to fertilization, or the punctures of the insect, or to both.

CĂPSŎMĀ'NĬĂ, an unnatural development of pistils. It may consist of excessive multiplication or of any alteration in form which impairs their function.

CĂP'SŪLAR, pertaining to a capsule.

CĂP'SŪLĀTE, enclosed in a capsule.

CĂP'SŪLE, any dry dehiscent fruit, especially one which is superior and polycarpellary. The sporangium of mosses is usually called a capsule.

CĂPSŪLĬF'EROŬS, bearing capsules.

CÄRBŎNĀ'ÇEOŬS, appearing as if burnt.

CÄR'ÇERŪLE, a dry fruit formed from a polycarpellary ovary, the carpels of which separate when ripe into indehiscent few-seeded cocci, as mallow; carcerulus. Compare SCHIZOCARP.

CARÇĔR'ŬLŬS, see CARCERULE.

CARÇĬTH'ĬUM, an old term for Mycelium.

CĂRĬCŎL'ŌGY̆, the study of the genus Carex.

CĀ'RĬĒȘ, an old term for decay.

CĂRĪ'NĂ, see KEEL.

CĂRĪ'NĂL, pertaining to a keel.

CĂRĪ'NĂL ǢSTĬVĀ'TION, when the carina embraces the other parts of the flower.

CĂRĪ'NĂL CĂNĂL', a lacuna in the xylem of a fibrovascular bundle, as in Equisetum. Compare VALLECULAR CANALS.

CĂR'ĬNĀTE, keel-shaped, or having a longitudinal ridge like a keel; keeled.

CĂR'ĬNĀTED, see CARINATE.

CĂRĬŎP'SĬS, see CARYOPSIS.

CĀ'RĬOŬS, decayed. (Rare.) Compare CARIES.

CÄR'NĒOŬS, flesh-colored; pale red. Compare CARNOSE.

CÄR'NĒŬS, see CARNEOUS.

CÄRNĬV'ŌROŬS, see INSECTIVOROUS.

CÄR'NŌSE, fleshy in texture. More firm than succulent or pulpy.

CÄR'NOŬS, pertaining to flesh; fleshy.

CÄRPĂDĒ'LĬŬM, see CREMOCARP.

CÄR'PEL, a simple pistil or one of the elements (modified leaves) of a compound pistil.

CÄR'PELLĀRY̆, pertaining to a carpel.

CÄRPĔL'LŬM (pl. **Cärpĕl'là**), see CARPEL.

CÄR'PĬD, a ripe carpel, especially if separable.

CÄRPĬD'ĬŬM, a carpid or carpel.

CÄRPŌĠĔN'ĬC, fruit-producing. In Florideæ applied to the special cell (or cells) of the procarpium which develops into the carpogonium.

CÄR'PŌGŎN, see CARPOGONIUM.

CÄRPŌGŌ'NĬŬM, the female organ, usually multicellular, of Carposporeæ before fertilization. Compare OOGONIUM.

CÄR'POLITE, a petrified fruit; lithocarp; carpolith.

CÄRPŎL'ŎĠY̆, the part of botany which relates to the structure of fruits. Compare POMOLOGY.

CÄR'PŌPHŌRE, the stalk of a sporocarp; a stalk elevating the gynecium above the rest of the flower (gynophore); a prolongation of the axis between the carpels, as in Umbelliferæ.

CÄR'PŌPHY̆LL, see CARPEL.

CÄRPŌPHY̆L'LŬM (pl. **Cärpōphy̆l'là**), see CARPEL.

CÄR'PŌSPŌRE, a spore produced in a sporocarp. De Bary extends the term to all those produced on a sporophyte, thus including the spores of ferns. This makes it equivalent to the term "spore" as used by Sachs.

CÄR'PŌSTŌME, the opening in a sporocarp through which the spores are discharged.

CÄRTĬLĂĠ'ĬNOŬS, firm and tough like cartilage.

CĂR'ŬNCLE, a partial aril growing from the hilum, as in Polygala; strophiole.

CĂRŬN'CŪLÀ, see CARUNCLE.

CĂRŬN'CŪLÄR, see CARUNCULATE.

CĂRŬN'CŪLĀTE, having a caruncle or the form of a caruncle; caruncular; carunculous; carunculated.

CĂRŬN'CŪLĀTED, see CARUNCULATE.

CĂRŬN'CŪLOŬS (rare), see CARUNCULATE.

CĂRY̆ŌÇĬNĒ'SĬS, see KARYOKINESIS.

CĂRY̆ŌPHY̆LLĀ'ÇEOŬS, pink-like; especially having five petals with long claws, as in the pink family (Caryophyllaceæ).

CĂRY̆ŎPH'Y̆LLOŬS, see CARYOPHYLLACEOUS.

CĂRY̆ŎP'SĬS, a dry one-seeded indehiscent fruit, with the pericarp thin and adherent to the seed, as in wheat and most other Gramineæ; cariopsis.

CĂSQUE, see GALEA.

CĂSSĬD'ĒOŬS, a term applied to an irregular corolla having the upper petal broad and helmet-shaped as in Aconitum.

CĂSTĀ'NĒOŬS, chestnut-colored.

CĂS'TRĀTE (adj.), said of a stamen which contains no anther, or no good pollen.

CĂS'TRĀTE (v.), to remove the anthers.

CĂTĂBŎL'ĬC, see KATABOLIC.

CĂTĂCLĒ'ŞĬŬM, a term sometimes applied to an achenium like Mirabilis, being a one-celled, one-seeded fruit within a hardened calyx.

CĂTĂCŌRŎL'LÀ, a secondary corolla sometimes found inside or outside the true corolla.

CĂTĂPĔT'ĂLOŬS, having the bases of the petals in a polypetallous corolla adherent to the bases of the stamens, as in Malva.

CĂT'ĂPHŸL, a scale-like leaf, as on buds, rhizomes, etc. Compare EUPHYLL.

CĂTĂPHŸL'LĂRŸ, scale-like, as the perules of a bud or the rudimentary leaves on a root-stock.

CĂT'ĔNĀTE, see CONCATENATE.

CĂTĔN'ŪLĀTE, see CONCATENATE.

CĂTH'ŌDĂL, see CATHODIC.

CĂTHŎD'ĬC, applied to the edge of a leaf which is entered by an ascending spiral. In a right-handed spiral it would be the left edge and *vice versa*. Also spelled Kathodic. Compare ANODIC.

CĂT'KĬN, see AMENT.

CẠU'DĀTE, having a long termination like a tail.

CẠU'DĀTED, see CAUDATE.

CẠU'DEX (pl. **Cạu'dexes** or **Cạu'diçēs**), the trunk of a palm, a tree-fern, or other arborescent endogen or acrogen; an upright root-stock; the persistent base of various herbaceous perennials (in this sense obsolete). Formerly applied to the trunk of any tree.

CẠU'DĬCLE, a little stalk, as that to which each pollen-mass in orchids is attached.

CẠUDĬC'ŪLĂ, see CAUDICLE.

CẠULĔS'ÇENT, having a distinct leaf-bearing stem.

CẠU'LĬCLE, the first internode, or portion of the stem below the cotyledons and above the radicle or beginning of the true root; tigellum; cauliculus, usually called radicle. Seldom applied to the part after the plant has developed.

CẠU'LĬCŪLE, see CAULICLE.

CẠULĬC'ŪLŬS (pl. **Cạulic'ūlī**), see CAULICLE.

CẠULĬF'ĔROŬS, having a stem; caulescent.

CẠU'LĬFÔRM, stem-like.

CẠULĬG'ĔROŬS, borne upon the stem.

CẠU'LĪNE, pertaining to the stem; belonging to the stem, or main stem, as *cauline leaves*.

CẠULĪNE BŬN'DLES, fibrovascular bundles confined exclusively to the stem and not connected with those in the leaves. Compare COMMON BUNDLES.

CẠU'LĬS, the stem; especially the main stem in herbaceous plants.

CẠULŌCĂR'POŬS, producing fruit upon a permanent stem, as ordinary trees and shrubs.

CẠU'LŌME, a general term for stems of all kinds, whether having the ordinary form and functions of stems or not. Compare PHYLLOME.

CẠULŎ'MER, a name proposed for one of the secondary axes which form a sympodium.

CẠULŌTĂX'ĬS, the arrangement of the branches upon a stem.

CĂV'ĬTŸ (Hort.). the depression in the stem end of an apple. Compare BASIN.

ÇĔLL, (1) one of the structural elements of living bodies, by the multiplication of which growth is effected. In plants the cell usually appears as a closed sac surrounded by a firm wall of cellulose and containing the essential element, protoplasm, and usually a nucleus, the active agent in cell-division. See CELLULOSE, CELL-WALL, CELL-SAP, PROTOPLASM, NUCLEUS. (2) Any cavity, as that of an anther or ovary.

ÇĔLL-FAMILY, a group of cells of common origin: used mainly among the lower algæ; colony. See CŒNOBIUM.

ÇĔLL-FĪ'BRE, see NUCLEAR FIBRIL.

ÇĔLLĬF'ĔROŬS, bearing or producing cells.

ÇĔLL-NŪ'CLĔŬS, see NUCLEUS.

ÇĔLL-PLATE, the early condition of a cell-wall at the equator of a dividing nucleus, formed by the fusion of a series of thickenings of the spindle-fibres called by Strasburger Dermatosomes.

ÇĔLL, PRĪMÔR'DĬAL, see PRIMORDIAL CELL.

ÇĔLL SĂP, fluid in living cells which separates from the protoplasm as one or more vacuoles; cytenchyma. Compare VASCULAR SAP and DEUTOPLASM.

ÇĔLL-TĬS'SŪE, tissue composed of cells as distinguished from vessels. See CELLULAR TISSUE.

ÇĔL'LŪLAR, composed of cells; pertaining to cells as distinguished from vessels; having the cells large and soft; containing cavities of any kind; porous.

ÇĔL'LŪLAR BÄRK, see MESOPHLŒUM.

ÇĔL'LŪLAR ĔN'VĔLŌPE, see MESOPHLŒUM.

ÇĔL'LŪLAR PLĂNTS, those which contain no fibrovascular tissue, as fungi and algæ; thallophytes.

ÇĔL'LŪLAR SPŌRE, see COMPOUND SPORE.

ÇĔL'LŪLAR SȲS'TĔM, the cellular portion of a plant as distinguished from the fibrovascular.

ÇĔL'LŪLAR TĬS'SŪE, tissue in which none of the cells are modified into ducts or vessels. Especially, tissue composed of cells which are large and loose, like the pulp of fruits; parenchyma. Compare VASCULAR TISSUE.

ÇĔL'LŪLE, a small cell.

ÇĔLLŬLĬF'ĔROŬS, bearing or producing little cells.

ÇĔL'LŪLŌSE, primary cell-wall substance; the material forming most of the dry matter of plants. It is seen nearly pure in cotton-fibre and in the cell-walls of most plants while young. There are many modifications of cellulose, among them Lignin in older wood and Suberin in bark.

ÇĔLL-WẠLL, a sac enclosing the living contents of a cell.

ÇĔMĔNTĀ'TION OF HȲ'PHÆ, their inseparable union by a cementing substance; concrescence. (DeBary.)

ÇĔM'ĔNT-DĬSK, Müller's term for the glandular disk or retinaculum of orchids.

ÇĔNĂN'THȲ, the suppression of stamens and pistils.

ÇĔNŌ'BĬŬM, see CŒNOBIUM.

ÇĔNŎĠENĔT'ĬC, secondary.

ÇĔN'TRĂL CĔLL, the cell of an archegonium, from which the oosphere originates. Compare EMBRYO-SAC.

ÇĔN'TRĂL CÔRD, a cord or bundle of elongated thin-walled cells at the centre of the stems, leaves, and fruit-stalks of many mosses which serves for the transfer of water; tissue-cord.

ÇĔN'TRĂL ÇȲL'ĬNDĔR, the portion of the tissue of roots and stems within the cortex. It includes the medullary and fibrovascular systems, — used mainly in roots. Compare FIBROVASCULAR CYLINDER.

ÇĔN'TRĬC, a term applied to such leaves as show no considerable difference between the internal structure of their upper and under sides. Compare BIFACIAL.

ÇĔNTRĬF'ŪGĂL, said of an inflorescence in which the terminal flower blossoms first; definite; determinate. Applied to a radicle which is turned toward the side of the fruit, or to anything else which points outward.

ÇĔNTRĬP'ĔTĂL, said of an inflorescence in which the lower or outer flowers blossom first; indefinite; indeterminate. Applied to radicles or seeds which point toward the axis of the fruit. Now little used in the latter sense.

ÇĔN'TŪRȲ, a hundred things, as a package of plants containing a hundred specimens.

ÇĔPHĂLĂN'THĬŬM, see ANTHODIUM.

ÇĔPHĂLŌ'DĬĂ (sing. **Cĕphălo'dĭŭm**), outgrowths of a lichen thallus in which algal cells are situated.

ÇĔPH'ĂLOID, capitate or head-shaped.

ÇĒRĀ'ÇEOŬS, wax-like.

ÇĒRĂMĬD'ĬŬM (pl. **Cĕrămĭd'ĭă**), the ovate or urn-shaped capsule containing the spores in the red algæ (Florideæ).

ÇĒRĬF'EROŬS, producing wax.

ÇĔR'NOŬS, nodding, curved over near the top, as the flower of narcissus upon its stem; nutant. Compare DROOPING.

ÇĔR'NŬOŬS, see CERNOUS.

ÇĔR'NŬŬS, see CERNOUS.

ÇĒ'SIOŬS, see CÆSIOUS.

ÇĔS'PĬTŌSE, in tufts or dense bunches; cæspitose; tufted.

ÇĔS'PĬTOŬS, see CESPITOSE.

ÇĔSPĬT'ŪLŌSE, in small tufts or bunches.

CHÆ'TĂ, Greek for bristle.

CHĂFF, the glumes and palets of grains and other grasses; the bracts which subtend each flower in the head of Compositæ, etc.

CHĀIN-GĔM'MĂ, a kind of gemma found in Mucorini having the form of a septate confervoid filament, the segments of which are capable of germination; sprouting gemma. (DeBary.)

CHĂLĀ'ZĂ, the base of the ovule or place where its coats unite with each other and with the nucleus. In orthotropous seeds it corresponds with the hilum.

CHĂP'LĔT, a series of objects arranged like beads on a string, as the spores of Cystopus.

CHĂR'ĂCTĔR, whatever distinguishes a plant or group of plants from others; a description composed of the distinctive features of a species or other group. See SPECIFIC CHARACTER, GENERIC CHARACTER, etc.

CHÄRTĀ'ÇEOŬS, of the texture of parchment or writing-paper.

CHĂSMŌG'ĂMȲ, the opening of the perianth at flowering time. (Rare.)

CHĬNK'Ȳ (Tuckerman), see RIMOSE.

CHLĂMY̆'DŌSPŌRE, a kind of thick-walled spore formed asexually in Mucorini by free-cell formation within the hyphæ. The term is also applied to certain thick-walled spores in protophytes, and occasionally elsewhere.

CHLŌ'RĂNTHY̆, the conversion of the parts of a flower into green leaves.

CHLŌRŌLEŪ'CĪTE, see CHLOROPLAST.

CHLŌ'RŌPHY̆L, see CHLOROPHYLL.

CHLŌ'RŌPHY̆LL, the green coloring matter of plants.

CHLŌ'RŌPHY̆LL-BŌD'Y̆, a proteid body (plastid) in the cells of plants containing the chlorophyll; chlorophyll-granule; chlorophyll-corpuscle; chloroplast; autoplast; chloroleucite. See PLASTID and CHROMATOPHORE.

CHLŌ'RŌPHY̆LL - CÔR'PŬSÇLE, see CHLOROPHYLL-BODY.

CHLŌ'RŌPHY̆LL-GRĀIN, see CHLOROPHYLL-BODY.

CHLŌ'RŌPHY̆LL-GRĂN'ŪLE, see CHLOROPHYLL-BODY.

CHLŌ'RŌPHY̆LL-VĔS'ĬCLES, see PYRENOIDS.

CHLŌ'RŌPLĂST, see CHLOROPHYLL-BODY.

CHLŌRŌ'SĬS, an abnormal absence or diminution of the green coloring matter of plants; albinism. Compare ETIOLATION.

CHÔR'DȦ PĬSTĬLLĀ'RĬS, a fibrovascular bundle in the style. (Obs.)

CHŌRĬPĔT'ȦLOŬS, see POLYPETALOUS.

CHŌRĬSĔP'ȦLOŬS, see POLYSEPALOUS.

CHŌ'RĬSĬS, the production of two or more organs in the position of one. It is considered in most cases to be a branching of very early origin. The condition may be normal or abnormal. See COLLATERAL and TRANSVERSE CHORISIS. Compare UNLINING.

CHŌRĬSTŎPH'Y̆LLOŬS, separate-leaved. (Rare.)

CHRŌMĂT'ĬC, capable of being colored by staining agents; pertaining to color.

CHRŌ'MȦTĬN, that part of the protoplasm of the cell, mainly in the nucleus, which is readily and deeply colored by staining agents. It forms the chief portion of the granular fibrils called "nuclein" or "nucleoplasm."

CHRŌMȦT'ŌPHŌRE, the protoplasmic body (plastid) in which chlorophyll or other coloring matter is produced; color-granule; chromoplast; chromoleucite. The term is also extended by some to all plastids. See CHLOROPHYLL-BODY.

CHRŌMĬD'ĬŬM, a term sometimes applied to the gonidium or algal host of lichens.

CHRŌMŌLEŪ'ÇĪTE, see CHROMATOPHORE.

CHRŌ'MŌPLĂST, see CHROMATOPHORE.

CHRŌMŌPLĂS'TĬD, see CHROMATOPHORE.

CHRŌ'MŌSŌME, one of the segments of the nuclear filament in karyokinesis.

CHRŎM'ŪLE, a term applied to all coloring pigments found in living plants.

CHRŎN'ĬZŌŌSPŌRE, one of the microzoogonidia produced in vast numbers in Hydrodictyon —so called, because they rest for several weeks or more before germinating; chronispore; chronizoospore.

ÇĬC'ĀTRĬÇE, see CICATRIX.

ÇĬCĀ'TRĬX (pl. **Çĭcătrī'çēs**), a scar left by a falling leaf or other organ.

ÇIĔN'ͼHȲMȦ, a system of intercellular spaces. (Kœhler.)

ÇĬL'ĬȦ (sing. **Çĭl'ĭŭm**), hairs or bristles forming a row or fringe; the tail-like appendages of zoospores; single, fine, soft, hair-like terminal appendages of any kind. Compare FLAGELLA.

ÇĬL'ĬĀTE, having cilia.

ÇIL'IATE-DĔN'TĀTE, having teeth fringed with hairs.

ÇĬL'ĬĬFÔRM, resembling cilia.

ÇĬL'ĬŎGRĀDE, moving by means of cilia, as zoospores.

ÇĬLĪ'ŎLȦ (sing. **Çĭlī'ŏlŭm**), secondary or minute cilia.

ÇĬL'ĬŬM (sing.), see CILIA.

ÇĬN'ÇĬNNĂL ÇȲME, see SCORPIOID CYME.

ÇĬN'ÇĬNNĂL DĪͼHŎT'ŌMȲ, see SCORPIOID DICHOTOMY.

ÇĬN'ÇĬNNŬS, see SCORPIOID CYME

ÇĬNĒRĀ'ÇEOŬS, a little paler than Cinereous.

ÇĬNĒ'RĒOŬS, ash-gray.

ÇĬNĒRĔS'ÇENT, becoming ash-gray; cineraceous.

ÇĬNĒRĬ'TIOŬS, see CINEREOUS.

ÇĪ'ŎN, see SCION.

ÇĬR'ÇĬNAL, see CIRCINATE.

ÇĬR'ÇĬNĀTE, rolled forward from the end, as the young leaves of many ferns.

ÇĬR'CŬMÇĪŞED', divided in a transverse circular manner.

ÇĬRCŬMNŪTĀ'TIŎN, the act of bending around successively through different points of the compass, as is done by the ends of twining vines, and in a less degree by other growing points; revolving nutation. See NUTATION.

ÇĬRCŬMSÇĬS'SĬLE, dehiscing or dividing by a transverse circular line, as the fruit of purslane.

ÇĬRCŬMSCRĬP'TION, an outline or boundary.

ÇĬRRHĬF'ĔROŬS, having tendrils. See CIRRHOSE.

ÇĬR'RHĬFÔRM, tendril-shaped; cirriform.

ÇĬR'RHŌSE, having tendrils; cirrhiferous; cirrhous; cirrous; cirrose; capreolate. Also, resembling a tendril or coiling like a tendril; tipped with a wavy filiform appendage; sleuderly flagellate.

ÇĬR'RHŌSELȲ PĬN'NĀTE, pinnate with a terminal leaflet replaced by a tendril, as in the pea.

ÇĬR'RHOŬS, see CIRRHOSE.

ÇĬR'RHŬS (pl. **Çĭr'rhī**), a tendril; capreolus.

ÇĬRRĬF'ĔROŬS, see CIRRHIFEROUS.

ÇĬR'RĬFÔRM, see CIRRHIFORM.

ÇĬRRĬG'ĔROŬS, producing tendrils. See CIRRHOSE.

ÇĬR'RŌSE, see CIRRHOSE.

ÇĬR'ROŬS, see CIRRHOSE.

ÇĬR'RŬS (pl. **Çĭr'rī**), see CIRRHUS.

ÇĬSTŌ'MȦ, see CISTOME.

ÇĬS'TŌME', a term sometimes given to a stomatic chamber when it exists as a kind of sac lined by a special layer of cells.

CLĂDOCĂR'POŬS, having the fruit in mosses on short lateral branches; pleurocarpous. Compare ACROCARPOUS.

CLĂD'ŌDE, see PHYLLOCLADIUM.

CLĂDŌ'DĬŬM (pl. **Clădōd'ĭȧ**), see PHYLLOCLADIUM.

CLĂD'ŌPHȲLL, see PHYLLOCLADIUM.

CLĂMP-ÇĔLL, see CLAMP-CONNECTION.

CLĂMP-CŌNNĔC'TION, a semicircular process connecting two adjoining cells of a hypha. In some cases it may communicate with but one of the cells, that from which it originated, in others its extremity becomes applied to the wall of the adjoining cell, at which point both walls are absorbed, leaving a passage around the septum between the cells of the filament. The process may become cut off from the originating cell by a septum when it is called a "clamp-cell." It is found mainly in Basidiomycetes.

CLȦSS, a group of related orders.

CLȦSSĬFĬCĀ'TION, the systematic distribution of individuals into groups.

CLĂTH'RĀTE, latticed; furnished with openings like lattice-work; cancellate.

CLĂTH'RĀTE ÇĔLL, see SIEVE-TUBE.

CLĀ'VĀTE, club-shaped; gradually thickened upward or away from the extremity of attachment.

CLĂV'ÆFÔRM, see CLAVATE.

CLĂV'ĔLLĀTE, diminutive of Clavate.

CLĂV'ĬCLE, see TENDRIL.

CLĂVĬC'ŪLȦ (pl. **Clăvĭc'ūlæ**), a term formerly used for tendril, especially a tendril formed by a petiole.

CLĂVĬC'ŬLĀTE, having a tendril. (Rare.)

CLĂV'ĬFÔRM, see CLAVATE.

CLĂV'ŪLE, the club-shaped conjugating process in Mucorini. (DeBary.)

CLĄW, the stem or narrow base of a petal, as those of the pink; unguis.

CLĔFT, having narrow sinuses reaching about half way to the base. Compare PARTED and DIVIDED.

CLEĪ'STŎCÄRP, an ascocarp which is entirely closed, and from which the spores escape by its final rupture, as that of the Erysipheæ.

CLEĪSTŎG'ĂMOŬS, having closed fertilization: a term applied to certain more or less depauperate flowers, sometimes underground, which never open, but are self-fertilized, as in some violets; clandestine. Cleistogamous flowers sometimes accompany flowers of the usual form on the same plant.

CLEĪSTŎG'ĂMȲ, self-fertilization without the flowers opening; closed-fertilization.

CLEĪ'STŎĠĒNE, a plant which bears cleistogamous flowers. See above. It may bear flowers of the ordinary form also.

CLĔS'TĪNE, see RAPHIDIAN CELL.

CLĬMB'ĬNG, rising by laying hold of other objects in any way except by twining; scandent.

CLĬNĂN'DRĬŬM, the part of the column of orchids containing the anther.

CLĬNĂN'THĬŬM, the receptacle of the flowers in Compositæ; phoranthium; clinium; anthoclinium. Compare ANTHODIUM.

CLĬNG'STŌNE, applied to varieties of the peach and other drupaceous fruits, in which the flesh when ripe is not readily separable from the pit. Compare FREESTONE.

CLĬNĬD'ĬŬM, the stalk or filament

supporting a stylospore. (Tuckerman.)

CLĬN'ĬŬM, see CLINANTHIUM.

CLĬNŌSPŌRĂN'ĠĬŬM, see PYCNIDIUM.

CLĬ'NŌSPŌRE, see STYLOSPORE.

CLĪSTŌ-, see CLEISTO-.

CLŌSED, applied to fibro-vascular bundles in which all the cambium cells become permanent tissue, as in monocotyledons. Compare OPEN.

CLŌSED - FĔRTĬLĬZĀ'TION, see CLEISTOGAMY.

CLŌSE-FĔRTĬLĬZĀ'TION, the fertilization of pistils by pollen from the same flower; self-fertilization.

CLŌS'ĬNG-MĔM'BRĀNE, the original unthickened cell-wall at the centre of a bordered pit.

CLOU'DĔD, having a pale ground with ill-defined patches of a darker tint gradually shading into it.

CLŌVE, one of the small bulbs developed in the axils of the scales of a larger bulb, as in garlic.

CLŬB, one of the elements composing the pulp in the seed-cavities of the lemon and orange. It consists of a more or less oval body about one fourth of an inch long filled with juice and attached by a stalk to the inner side of the cavity or cell. Morphologically, it is a pluricellular hair.

CLŬBBED (Hort.), applied to the stem of an apple when its base is enlarged and fleshy. Compare CLAVATE.

CLŬB-SHĀPED, see CLAVATE.

CLŬS'TĔR, an indefinite popular term, applied mainly to small-fruits in which several fruits grow together in an inflorescence, as in the raceme of the currant. About the same as Bunch, but applied more often perhaps to loose collections.

CLŬS'TĔR-CŬP, see ÆCIDIUM.

CLŬS'TERED, collected into or growing in a bunch, as the conglomerate flowers of dodder.

CLȲP'ĒATE, see SCUTATE.

CLȲ'PĔĬFŌRM, see SCUTATE.

CŌĂÇĔR'VĀTE, see ACERVATE.

CŌĂD'ŬNĀTE, united at the base or farther; connate.

CŌAĒTĀ'NĒOŬS, existing or appearing at the same time; contemporaneous.

CŌALĔS'ÇENÇE, the complete union of similar things. The same as Cohesion when applied to organs. Compare CONJUGATION.

CŌALĔS'ÇENÇE OF ÇĔLLS, the partial or entire absorption of the partition-walls of adjoining cells, as when the cavities of long rows of cells in a tissue become connected in the formation of ducts or vessels.

CŌĂRC'TĀTE, crowded together; compact. Opposed to effuse.

CŌAT'ĔD, composed of layers, or having a rind.

CŌATĔD BŬLB, see TUNICATED BULB.

CŎB'WĔBBȲ, having fine slender filaments or hairs like cobweb; arachnoid.

CŎÇÇĬD'ĬŬM (pl. **Cŏççĭd'ĭà**), a sporocarp like that of Delesseria (one of the algæ), being a closed case with the spores borne on a central placenta.

CŎÇÇĬF'ĔROŬS, producing berries; bacciferous.

CŎÇÇĬN'ĔŬS, scarlet.

CŎC'CŬS (pl. **Cŏc'çī**). (1) One of the separable one-seeded carpels of certain dry fruits, as Euphorbia. Compare REGMA. (2) A spherical or spheroidal cell among bacteria, especially of the genus Micrococcus.

CŌCH'LĔĂ, see STROMBUS.

CŌCH'LĔĂR, a form of æstivation in which one piece being larger covers all the others, as in Aconitum. Compare VEXILLARY.

CŌCHLĔĂR'ĬFŌRM, shaped like the bowl of a spoon.

CŌCH'LĔĀTE, shaped like a snail-shell; spirally turbinate. Practically the same as Strombuliform.

CŎCKS'CŌMBED, fasciated. Applied mainly in horticulture to strawberries which are irregular in shape from being flattened at the sides or fasciated.

ÇŒL'ŌSPĔRM, a cœlospermous seed.

ÇŒLŌSPĔR'MOŬS, applied to the seeds (cremocarps) of coriander and some other Umbelliferæ, which have the inner surface hollowed by the curving in of the top and bottom.

ÇŒNĂN'THĬŬM, see CLINANTHIUM. Also sometimes applied to receptacles like the fig, or to any fleshy receptacle in which the flowers are more or less embedded.

ÇŒNŌ'BĬŬM (1) a community of a definite number of unicellular individuals united in one body of definite form, as in Volvox: cenobium. Compare CELL-FAMILY. (2) A name of the fruit peculiar to the Boraginaceæ and Labiateæ, consisting of four distinct nutlets around a common style.

ÇŒRULĔS'CĔNT, bluish, lighter than cœruleus.

ÇŒRU'LĔŬS, light blue; sky-blue.

CŌĔTĀ'NĔOŬS, appearing or existing at the same time; of the same age.

CŌHĒ'RENT, having similar parts more or less united, as the petals in a gamopetalous corolla. Compare ADHERENT.

CŌHĒ'ŞION, the union of members of the same whorl or kind. Compare ADHESION.

CŌ'HŌRT, a natural group of orders within a class; alliance.

CŌLĔŌPHȲL'LŬM, a membranous or fleshy sheath investing the plumule in Monocotyledons.

CŌLĔŌP'TĬLE, see COLEOPHYLLUM.

CŌLĔŌRHĪ'ZĂ, the covering through which the radicle of most Monocotyledons bursts in germination; root-sheath. The term has also been applied to the vascular-bundle sheath in roots.

CŌL'LAR, the line of junction between the stem and root; collum; neck. Also applied to the annulus of a mushroom.

CŌLLĂTĔRAL, side by side.

CŌLLĂT'ĔRAL BŬN'DLE, a fibrovascular bundle having a single strand of phloem in continuous contact with a single strand of xylem. Compare BICOLLATERAL BUNDLE, RADIAL BUNDLE, and CONCENTRIC BUNDLE.

CŌLLĂTĔRAL CHŌ'RĬSĬS, when the parts originating by chorisis stand side by side; parallel chorisis. Compare TRANSVERSE CHORISIS.

CŌLLĔCT'ĬNG HÂIRS, hairs upon the style in certain Compositæ which serve to collect the pollen as it is discharged from the anther.

CŌLLĔCT'ĬVE FRUIT, a fruit-like body originating from more than one flower, as the mulberry; multiple fruit; pseudo-syncarp. Compare ANTHOCARPOUS FRUIT, ACCESSORY FRUIT, and AGGREGATE FRUIT.

CŌLLĔCT'ORS, see COLLECTING HAIRS.

CŎLLĔN'CHȲMȦ, tissue composed of cells having cartilaginous thickenings at the angles. Common beneath the epidermis of stems, and in other places where strength is required.

CŌL'LĔT, an old term for collar.

CŎLLĒ'TĒRS, glandular hairs which secrete a gummy matter (blastocolla) upon buds.

CŌL'LŌID, n., any substance in the colloid state.

CŌL'LŌID, adj., resembling jelly; the molecular condition of a class of substances, such as gums, usually produced by the disorganization of organized matter, which mix with water in all proportions and pass from the solid to the fluid state through all stages of softening, thus differing from Crystalloids, which see.

CŌLLŌID'AL, see COLLOID.

CŌL'LŬM, the neck or tapering base of the capsule in mosses; the line of junction between root and stem; collar.

CŌL'MAR SHĀPED, pear-shaped with a rather slender neck and large body. (J. J. Thomas.)

CŎL'ŌNȲ, see CELL-FAMILY.

COL'ŌRED, of any other color than green.

CŌLPĔN'CHȲMȦ, epidermal tissue composed of cells with sinuous margins. Not in general use.

CŌLŬMĔL'LȦ, the axis or central column of a pod or spore-case. Usually a placenta.

CŌLŬMĔL'LĬFŌRM, shaped like a little column or columella.

CŌL'ŬMN, the united filaments and styles in a gynandrous flower, as an orchid.

CŌLŬM'NAR, shaped like a column or pillar: round or nearly so, tapering slightly or none, and not so long as to be called slender. Often applied to styles. Compare TERETE.

CŌ'MȦ, a tuft of hairs on a seed; a terminal cluster of empty bracts; any tuft.

CŌ'MĀTE, see COMOSE.

CŌMBĪNED' HȲ'BRĬD, a derivative hybrid in which three or more species or varieties are united, as when a hybrid unites with a new parent form or another hybrid.

CŌMB-SHĀPED, see PECTINATE.

CŌMMĔN'SALĬSM, see SYMBIOSIS.

CŌM'MĬSSŪRE, a line of junction of two parts. In the Marattiaceæ a longitudinal partition connecting the two stipules and forming an anterior and a posterior chamber. Compare SUTURE.

CŌM'MŌN, general, primary, or universal, as opposed to partial or secondary.

CŌM'MŌN BŬD, one containing both leaves and flowers, or one from which more than one flower is produced.

CŌM'MŌN BŬN'DLE, a fibrovascular bundle a part or the whole of which passes from the stem into a leaf.

CŌM'MŌN ĬN'VŌLŪCRE, one subtending an inflorescence; general or universal involucre.

CŌM'MŌN NĀME, any name, except the "botanical name," by which a plant is known.

CŌM'MŌN PĔDŬN'CLE, one supporting several pedicels.

CŌM'MŌN PĔR'ĬĂNTH, applied to certain common involucres, as that surrounding the head of flowers in Compositæ.

CŌM'MŌN RĒÇĔP'TĂCLE, one supporting more than one flower or other organ.

CŌM'MŌN ŬM'BĔL, see COMPOUND UMBEL and UNIVERSAL UMBEL.

CŌ'MŌSE, bearing a coma or tuft, or growing in tufts; comate.

CŌMPĂCT', solid; close.

CŌM'PLĂNĀTE, flattened vertically to a level surface above and below. Applied also to several organs which are flattened or arranged so as to lie in one plane, and to such as lie flat upon or against each other, as when leaves lie flat upon the stem. Compare EXPLANATE and COMPRESSED.

CŌMPLĔTE', having calyx, corolla, stamens, and pistils. Compare PERFECT.

CŌM'PLĔX, an assemblage of interwoven fibres, or any group of complicated parts.

CŌM'PLĬCĀTE, folded together forward. Compare REPLICATE and CONDUPLICATE.

CŎM'POUND, consisting of a number of similar subordinate parts forming a complete whole, as the leaf of the walnut.

CŌM'POUND CŎR'ȲMB, one with more than one flower on each ray or branch.

CŌM'POUND DĪCHĀ'SĬŬM, one in which the primary axis terminates in a flower, beneath which arise several secondary axes, each of which terminates in like manner, as in valerian.

CŌM'POUND FLOW'ĔR, an old term for the flower-head in Compositæ.

CŌM'POUND FRUIT, see AGGREGATE FRUIT.

CŌM'POUND ĬN'FLŌRĔS'ÇENÇE, one in which the ultimate branches each bear more than one flower, thus forming a compound spike, corymb, etc.

CŌM'POUND LĔAF, one having two or more distinct blades, or leaflets, as in the ash.

CŌM'POUND Ō'VĂRȲ, one having more than one carpel.

CŌM'POUND PĬS'TĬL, one composed of more than one carpel.

CŌM'POUND RĂÇĒME', see PANICLE.

CŌM'POUND SPĪKE, one having more than one flower or spikelet on each short branch, as wheat. Applied especially when the secondary spikes are well developed, or when, as in *Panicum sanguinale*, the inflorescence consists of several spikes of nearly equal size arising from the apex of the peduncle.

CŌM'POUND SPŌRE, a spore consisting of more than one cell, each of which is frequently capable of germination. It differs from Gemma by its more definite form and specialized method of production. The synonyms are spore-group, semen-multiplex, compound spore, cellular spore, multicellular spore, pluricellular spore, septate spore, polyspore, sporidesm, etc. See MERISPORE.

CŌM'POUND SPŌ'RŌPHŌRE, one formed by the cohesion of separate hyphal branches. Compare SIMPLE SPOROPHORE.

CŌM'POUND STĚM, a branched stem.

CŌM'POUND ŬM'BĔL, one in which each primary ray bears a smaller umbel instead of a single flower. Formerly known as Common, General, or Universal Umbel.

CŌMPRĔSSED', flattened, especially lengthwise laterally, the horizontal diameter much less than the vertical. Compare DEPRESSED, OBCOMPRESSED, and COMPLANATE.

CŌNCĂT'ĔNĀTE, joined in a continuous series like a chain; catenate; catenulate.

CŌNCĂT'ĔNĀTĔD, see CONCATENATE.

CŌNCAULĔS'ÇENÇE, the coalescence of the pedicel of a flower with the stem for some distance above the subtending bract.

CŌNÇĔN'TRĬC, having a common centre.

CŎNÇĔN'TRĬC BŬN'DLE, a fibrovascular bundle in which a strand of one element is wholly surrounded by the other, as in some ferns in which the xylem is wholly surrounded by phloem.

CŌNÇĔP'TÀCLE, a name applied to sac-like receptacles of various kinds, as perithecium, cystocarp, follicle.

CŌNÇĔPTĂC'ŪLŬM, see CONCEPTACLE.

CONCH'ĬFŌRM, like half a clamshell.

CŌN'COLŌR, of the same color as another part or plant. Also applied to several objects which are all of the same color. Compare UNICOLOR.

CŎN'COLŎRED, see CONCOLOR.

CŌN'CŌLŌROŬS, see CONCOLOR.

CŌNCŌM'ĬTANT, applied by De Bary to fibrovascular bundles which run continuously side by side without becoming separated by other bundles.

CŎNCRĔS'ÇENÇE, see CEMENTATION.

CŌN'CRĒTE, grown together.

CŌNDŬCT'ĬNG ÇĔLLS, narrow elongated cells associated with sieve tubes, and similar to them, but without perforated walls.

CŌNDŬCT'ĬNG TĬS'SŪE, tissue composed of conducting cells. Compare CONDUCTIVE TISSUE.

CŌNDŬCT'ĬVE TĬS'SŪE, that through which the pollen tube passes on its way to the ovary. It is often loose in texture and moist with nutritive fluid for the growth of the pollen tube.

CŌNDŬ'PLĬCĀTE, folded together forward and lengthwise in any manner; complicate. Compare REPLICATE and INFLEXED.

CŌNE, the fruit of Coniferæ; strobile; also applied in other plants to a fruit or inflorescence resembling the cone of the fir or pine. See GALBULUS.

CŌNFĔRRU'MĬNĀTE, closely united, as the cotyledons of the horse-chestnut.

CŌNFĔR'TĔD, crowded or clustered; opposed to distinct. Compare CONGESTED.

CŌNFĔR'VŌID, loose and filamentose, like conferva among algæ.

CŎN'FLŪENT, running together or blended into one; coherent.

CŌN'FLŪENT FRUIT, an old term for collective fruit.

CŌNFŌRMED', (1) closely resembling; (2) closely fitted to, as the skin to a seed.

CŌN'ĠĔNĔR, a plant of the same genus as another.

CŌNĠĔNĔR'ĬC, said of two species or individuals which belong to the same genus.

CŎNĠE'RĬĒS, a collection of parts or organs. (Rare.)

CŌNĠĔST'ĔD, packed closely together.

CŌNGLŌ'BĀTE, clustered into a ball.

CŌNGLŌM'ĔRĀTE, clustered together; opposed to Diffuse.

CŌNGLŪ'TĬNĀTE, glued together in a mass.

CŌN'ĬCAL, in the form of a geometric cone. Said of a root when it tapers downward, or of any other part when it tapers upward or outward. Compare OBCONICAL.

CŌNĬD'ĬĀ, pl., see CONIDIUM.

CONIDIIF'ĒROŪS, bearing conidia.

CŌNĬD'ĬŌPHŌRE, a branch of mycelium bearing one or more conidia.

CŌNĬD'ĬŌSPŌRE, see CONIDIUM.

CŌNĬD'ĬŬM (pl. Cōnĭd'ĭà), an asexual spore (gonidium) abstricted singly or otherwise from the apex of a filament, as in Peronospora. Applied chiefly where the spores form a dusty or powdery coat, but not used in Hymenomycetes, and in less use elsewhere than formerly, the general term Gonidium being now usually employed instead.

CŌNĬF'ĔROŪS, cone-bearing.

CŌN'ĬFŌRM, see CONICAL.

CŌ'NĬŌÇȲST, a term applied by Harvey to the oogonium of Vaucherieæ.

CŌN'JŪGĀTE, joined or arranged in pairs, as the leaflets of many compound leaves; paired. See BINATE.

CŌNJŬGĀ'TION, the simplest method of fertilization, in which the male and female cells are alike or nearly so, as in the order Conjugateæ; zygosis. The uniting cells are called gametes and the product a zygote. The conjugation of naked protoplasmic bodies, as zoospores, is sometimes called Coalescence.

CŌNJŬGĀ'TION-ÇĔLL, see GAMETE.

CŌNJŬNC'TĬVE THRĔADS, Fol's term for spindle-fibres, which see.

CŌNNĂS'ÇENT, produced at the same time.

CŌN'NĀTE, said of organs of the same nature which are grown together from the first or united at the base. Compare ADNATE.

CŌN'NĀTE-PĔRFŌ'LĬĀTE, said of a pair of opposite leaves when their bases are united around the stem.

CŌNNĔCT'ĬNG CĔLL, Harvey's term for Heterocyst, which see.

CŌNNĔCT'ĬVE, the part of the anther (being a continuation of the filament) which connects its two lobes.

CŎNNĔCTĪ'VŬM, see CONNECTIVE.

CŌNNĪ'VENT, convergiug.

CŌNNŪ'BĬŬM, a term applied by Pringsheim to that stage in the conjugation of Conjugateæ in which the protoplasm of the conjugating cells has coalesced.

CŌ'NOID, cone-like; conical.

CŌNOID'AL, somewhat conoid.

CŌNSŌL'ĬDĀTĔD, (1) grown together, said either of like or unlike parts; (2) having a small surface in proportion to bulk, as many cacti.

CŎN'SÔRTĬSM, see SYMBIOSIS.

CŌN'STANT, always present, or always in the same condition; uniform.

CŎNSTRĬCT'ĔD, narrowed in certain places.

CŎNSTRŬCTĬVE MĒTĂB'ŌLĬSM, see ASSIMILATION.

CŎNTĂBĔS'ÇENÇE, the condition of being wasted away or aborted; said of anthers which contain little or no fertile pollen.

CŎNTĀ'GIOŬS, said of diseases which are communicable from one plant or animal to another by contact only, or by the direct transfer of the disease-producing organism. Strict contagion implies parasitism, the organism being unable to grow outside the supporting body. Compare INFECTIOUS.

CŎNTĔR'MĬNOŬS, of equal extent.

CŎNTĬG'ŪOŬS, near, or in contact.

CŎNTĬN'ŪOŬS, uniform in structure or outline; uninterrupted. Said of hyphæ which are without septa, or of objects which are in all parts of the same size, or whose diameter increases or diminishes regularly.

CŎNTÔRT'ĔD, in æstivation, when the margins of the floral leaves successively overlap each other (obliquely or otherwise) in one direction; twisted. Compare CONVOLUTE.

CŎNTÔRT'ĔD ÆSTĬVĀ'TION, see above.

CŎNTÔR'TION, an abnormal twisting of branches or other organs.

CŎNTÔRTŪ'PLĬCĀTE, twisted and folded.

CŎNTRĂCT'ĔD, narrowed, or the successive parts shortened. Compare CONSTRICTED.

CŎNTRĂCT'ĬLE VĂC'ŪŌLEŞ, small cavities containing a watery fluid which make their appearance in the protoplasm of many zoospores and other motile organisms and then suddenly disappear; pulsating vacuoles. Their function is not known.

CŎNTRĂCTĬL'ĬTY̆, a property of protoplasm by which it is enabled to change its form spontaneously, or by virtue of forces within itself. Compare IRRITABILITY.

CŎN'TRÀRY̆, extending in an opposite direction to something with which it is compared; as, the pod of shepherd's purse is flattened *contrary* to the partition.

CŌ'NŬS, see CONE.

CŎN'VŌLŪTE, rolled together lengthwise from one edge, as the leaves of the plum in the bud. The term Contorted (used mainly in æstivation) refers more especially to the relation of the organs to each other and to the axis, while Convoluted (used mainly in vernation) refers more to the manner of folding of the organ itself. A *contorted* corolla may or may not have its parts *convoluted*.

CŎN'VŌLŪTĔD, see CONVOLUTE.

CŎP'RŌPHȲTE, see SAPROPHYTE.

CŎR'ÀCOID, shaped like a crow's beak.

CŎR'ĂLLĬNE, resembling coral; coralliform; coralloid.

CÔR'CLE, see CORCULUM.

CÔR'CŪLŬM, an old term for

plumule, or plumule and radicle together.

CÔR'DĀTE, heart-shaped, as usually pictured, with the point of attachment at the broad end. Applied mainly to leaves.

CÔR'DĀTĔD, see CORDATE.

CÔR'DĀTE-HĂS'TĀTE, intermediate in form between cordate and hastate, approaching more nearly to hastate.

CÔR'DĀTE-Ō'VĀTE, between cordate and ovate, approaching more nearly to ovate.

CÔR'DĀTE - SĂĠ'ĬTTĀTE, between cordate and sagittate, approaching more nearly to sagittate.

CÔR'DĬFÔRM, sometimes applied to solid bodies shaped like the human heart. Compare CORDATE.

CŌRE, the bony endocarp of a pome containing the seeds.

CŌRĬĀ'ÇEOŬS, leathery in texture.

CŌRK, tissue composed of firm elastic cells, generally soon filled with air, which are produced in the bark and upon injured surfaces. It is composed of a modification of cellulose called Suberin, and is designed for protection.

CŌRK-CĂM'BĬŬM, see PHELLOGEN.

CÔRK-MĔR'ĬSTĔM, see PHELLOGEN.

CÔRK'Ÿ ĔN'VĔLŌPE, see CORKY LAYER.

CÔRKY LĀY'ĔR, the layer of bark immediately below the epidermis which produces the cork; epiphlœum. Compare MESOPHLŒUM.

CŌRM, the enlarged base of an herbaceous stem, consisting of one or a few short internodes, and serving for the storage of starch or other reserve food-materials; pseudo-bulb; solid bulb; bulbo-tuber. It differs from a Tuber mainly in being upright, or more nearly so, and in seldom being produced upon an elongated subterranean stem. The Indian turnip and crocus are examples. Compare PLATEAU.

CÔR'MŬS, see CORM.

CÔR'MŌPHŸTE, a plant having a true stem. Compare THALLOPHYTE.

CÔR'NĒOŬS, having the texture of horn.

CŎRNĬC'ŪLĀTE, having a process or appendage like a little spur or horn. Compare CORNUTE.

CÔR'NĬFÔRM, horn-shaped.

CÔR'NŪ (pl. Cŏr'nūȧ), see HORN.

CÔR'NŪTE, horn-shaped, as the nectary of columbine, or bearing a horn-shaped process; corniform. Compare CORNICULATE.

CŎR'ŎL (obs.), see COROLLA.

CŎRŎL'LȦ, the conspicuous part of most flowers, being the inner set of floral envelopes when there is more than one, commonly distinguished by its fine texture and by having some other color than green. See PETAL, CALYX, and PERIANTH.

CŎRŎLLĀ'ÇEOŬS, like a corolla in appearance or texture; petaloid. Petaloid is the term usually employed, especially when referring to individual organs.

CŎR'ŎLLĀTE, having a corolla.

CŎR'ŎLLĀTĔD, see COROLLATE.

CŎR'ŎLLĔT, see FLORET.

CŌRŎLLĬFLŌ'RAL, having calyx, petals, and ovary inserted separately on the receptacle and the stamens inserted upon the corolla. Compare CALYCIFLORAL and THALAMIFLORAL.

CŎR'ŎLLĪNE, see COROLLACEOUS.

CŎR'ŎLLŪLE, see FLORET.

CŌRŌ'NȦ (pl. **Cōrō'næ**), see CROWN.

CŌR'ŌNĀTE, having a corona or crown.

CŌRŎN'ĬFÔRM, having the form of a corona or crown.

CŌRŌ'NŪLȦ (pl. **Cōrō'nūlæ**), see CORONULE.

CŎR'ŌNŪLE, a diminutive corona or crown.

CÔR'PŬSÇLE, (1) any very small body; (2) see CORPUSCULUM (in the first sense).

CŎRPŬS'CŪLȦ, pl., see CORPUSCULUM.

CŎRPŬS'CŪLE, see CORPUSCULUM.

CŎRPŬS'CŪLAR, pertaining to or composed of corpuscles or small particles.

CŎRPŬS'CŪLŬM (pl. **Cŏrpŭs'cūlȧ**), (1) an old term for the central cell in the archegonium of Gymnosperms: sometimes applied to the whole archegonium; (2) the connecting body between the arms (retinacula) which bear the pollinia of milkweeds (Asclepias).

CÔR'ṚUGĀTĔD, in folds or wrinkles.

CÔR'TĔX, that portion of the fundamental tissue lying outside the fibrovascular bundles. In trees it is the "outer bark."

CÔR'TĬCAL, pertaining to the bark or cortex.

CÔR'TĬCAL SHĔATH, the ring of primary bast (phloem) bundles. Compare MEDULLARY SHEATH.

CÔR'TĬCĀTE, having a rind or cortex.

CÔRTĬÇĬF'ĔROŬS, producing bark or cortex.

CŎRTĬÇ'ĬFÔRM, like bark.

CŎRTĬC'ŌLOŬS, growing upon the bark of trees, as many mosses and lichens.

CŌR'TĬCŌSE, having or resembling bark; corticous.

CÔR'TĬCOŬS, see CORTICOSE.

CÔRTĪ'NȦ, the remnant of the veil sometimes found at the border of the pileus or cap in mushrooms.

CŌR'ȲMB, a convex or flat-topped indefinite or centripetal inflorescence, like a raceme with the lower pedicels elongated. Compare CYME.

CŌRȲM'BĀTE, having corymbs, or growing in corymbs.

CŎRȲMBĬF'ĔROŬS, producing corymbs.

CŌRȲM'BŌSE, growing in or resembling corymbs; corymbous.

CŌRȲM'BOŬS, see CORYMBOSE.

CŌRȲM'BŪLŌSE, arranged in small corymbs.

CŌRȲM'BŬS, see CORYMB.

CŎṢMŎP'ŌLĪTE, a plant widely extended in the world, as *Pteris aquilina*.

CŎS'TȦ (pl. **Cŏs'tæ**), see RIB.

CŎS'TAL, pertaining to a rib.

CŎS'TAL-NĔRVED, a term sometimes applied to parallel-veined leaves when the veins arise from a midrib and run toward the margins, as in the banana.

CŎS'TĀTE, having one or more prominent veins or ribs.

CŎSTĔL'LĀTE, having small ribs.

CŎTȲLĒ'DŎN, the first leaf or leaves of a plant; seminal-leaf; seed-lobe; seed-leaf. In exogens they are usually different in form from the leaves produced afterward, and are stored with nutritive matter for the young plant.

CŎTȲLĔD'ŎNOŬS, having cotyledons.

CŌTȲL'ĬFÔRM, like a cotyle, the articulating cavity of a bone; a little shallower than acetabuliform.

COWLED, see CUCULLATE.

CRĂM'PŎN, see HOLD-FAST.

CRÂTĔR'ĬFÔRM, goblet-shaped; narrower than Calathiform and less flaring than Cyathiform.

CREEP'ĬNG, running upon or under ground and rooting; repent.

CRĔM'ŎCÄRP, the fruit of Umbelliferæ. Also applied to other fruits of somewhat similar structure, as that of maple. See MERICARP and SCHIZOCARP.

CRĔ'NȦ, see CRENATURE.

CRĔ'NĀTE, having the margin furnished with rounded teeth, which usually point toward the apex of the object bearing them; crenated; crenelated; crenelled. See DOUBLY, ACUTELY, and OBTUSELY CRENATE.

CRĔ'NĀTĔD, see CRENATE.

CRĔN'ȦTŪRE, a tooth of a crenate margin; crenel; crena.

CRĔNĔL', see CRENATURE.

CRĔN'ĔLĀTĔD, see CRENATE.

CRĔNĔLED', see CRENATE.

CRĔNELLED', see CRENATE.

CRĔN'ŪLĀTE, diminutive of crenate; having the margin furnished with fine rounded teeth.

CRĔN'ŪLĀTĔD, see CRENULATE.

CRĔST, (1) a partial aril in the form of a ridge along the raphe of some seeds, as Sanguinaria and Dicentra. Compare ARIL, STROPHIOLE, and CARUNCLE. (2) An elevated ridge or appendage terminating any organ: a stamen is crested when the filament (connective) projects beyond the anther.

CRĔST'ĔD, having a ridge or terminal appendage; cristate.

CRĒTĀ'ÇEOŬS, (1) of the color of chalk; (2) growing on chalky land.

CRĬB'RĀTE, see CRIBROSE.

CRĬB'RĬFŌRM, see CRIBROSE.

CRĬB'RĬFŌRM ÇĔLLS, see SIEVE-TUBES.

CRĬB'RŌSE, pierced with small holes like a sieve; cribriform; cribrate.

CRĬB'RŌSE ÇĔLLS, see SIEVE-TUBES.

CRĬ'NĬTE, having a tuft or fringe of long weak hairs.

CRĬ'NOĪD, resembling a lily.

CRĬSP, having the surface, especially near the margin, strongly and finely undulate, as the leaves of Savoy cabbage or the cultivated endive; curled; crispate; crisped.

CRĬS'PĀTE, see CRISP.

CRĬSPED, see CRISP.

CRĬS'TĀTE, see CRESTED.

CRĬS'TĀTĔD, see CRESTED.

CRĬT'ĬCAL, difficult to characterize or classify.

CRŌSS, a union of two varieties of the same species. Applied in a narrower sense to the offspring of any two flowers which have been cross-fertilized. Compare HYBRID.

CRŎSS-ÄRMED, see BRACHIATE.

CRŎSS-BREED, see CROSS.

CRŎSS-FĔRTĬLĬZĀ'TION, the fertilization of a flower by pollen from another flower, especially from one of another variety of the same species; allogamy. Compare CLOSE-FERTILIZATION.

CRŎSS-PŎLLĬNĀ'TION, the conveyance of pollen to the stigma of another flower.

CROWN, (1) an appendage in the throat of the corolla in some flowers, as Silene and Narcissus; corona; paracorolla. In some cases it represents a circle of metamorphosed stamens. (2) A ring of cells at the apex of the nucule in Characeæ. (3) Any circle of organs in the form of a crown, as the scales at the apex of an achene.

CROWN'ĬNG, borne at the apex.

CRŌ'ZIĔR, anything with a coiled end, as the young leaves of most ferns.

CRU̥'ÇĬĀTE, in the form of a Maltese cross, as the petals of Cruciferæ; cruciform; cross-shaped.

CRU̥ÇĬF'ĒROŬS, (1) belonging to the Cruciferæ or mustard family; (2) resembling plants of the mustard family, especially in the form of the flower.

CRU̥'ÇĬFÔRM, see CRUCIATE.

CRŬMPLED ÆSTĬVĀ'TION, when the petals are irregularly folded in the bud, as in the poppy.

CRU̥'RAL, somewhat leg-shaped. Used mainly in composition.

CRŬSTĀ'ÇEOŬS, said of the thallus of a lichen when it forms an adherent crust which cannot be removed from the substratum without injury.

CRȲP'TŌGĂM, a plant belonging to the lowest of the two chief divisions of the vegetable kingdom, including ferns, mosses, fungi, etc. Cryptogams produce no true seeds or flowers, but in most groups there is a process of fertilization essentially the same as in the higher plants.

CRȲPTŎG'ÀMOŬS, pertaining to cryptogams, or plants having no true flowers or seeds.

CRȲP'TŌPHȲTE, see CRYPTOGAM.

CRȲS'TALLŌIDS, protein bodies in the form of crystals; albumen crystals. Applied also in opposition to Colloids to indicate all substances susceptible of crystallization and of diffusion through a membrane. (Graham.)

CŪ'CŬLLĀTE, conical, with the side cleft, and often inrolled like a cornet of paper, as the spathe of arum and the calyptra of some mosses; cuculiform; hooded; hood-shaped; cowled.

CŪCŬL'LĬFÔRM, see CUCULLATE.

CŪCŬL'LŬS (pl. **Cūcŭl'lī**), a term sometimes given to various hood-shaped organs, especially a concave and arched sepal or petal, as the large upper sepal (galea) of Aconitum.

CŪCŪ'MĬFÔRM, having the form of a cucumber.

CŪCÛRBĬTĀ'ÇEOŬS, belonging to or resembling plants of the Cucurbitaceæ or melon family.

CU̥L-DĔ-SĂC, a tubular or bag-shaped cavity closed at one end.

CŬLM, the stem of grasses; sometimes applied to that of sedges also.

CŬLMĬC'ŌLOŬS, growing upon the stems of grasses, as certain fungi.

CŬLMĬF'ĒROŬS, producing culms.

CŬL'TRĀTE, shaped like a pruning-knife; broad, thin, pointed, with the end curved edgewise; cultriform.

CŬL'TRĀTĔD, see CULTRATE.

CŬL'TRĬFÔRM, see CULTRATE.

CŪ'NĒAL, see CUNEATE.

CŪ'NĒĀTE, wedge-shaped in outline. Said of leaves which are broad above and narrowed to the base in straight lines. The same as Obdeltoid but usually narrower; cuneiform.

CŪNĒ'ĬFÔRM, see CUNEATE.

CŪNĬC'ŪLĀTE, pierced with a long narrow passage or passages.

CŬP, (1) a concave involucre enclosing a nut, as in the acorn; cupule; (2) a discocarp (apothecium) in Ascomycetes.

CŬP-SHĀPED, a rather indefinite term, usually applied to cavities from one half to twice as deep as broad, with concave bottom and nearly vertical sides

CŪ'PŪLÀ, see CUP.

CŪ'PŪLĀTE, furnished with a cup or cupule.

CŪ'PŪLE, see CUP.

CŪPŪLĬF'ĒROŬS, bearing or furnished with a cupule or cupules.

CŪ'PŪLĬFÔRM, see CUP-SHAPED.

CÛRD, applied in horticulture to the material composing the head in cauliflower. Sometimes the heads individually are called "curds."

CÛRLED, see CRISP.

CÛRVĬCAU'DĀTE, having a curved tail.

CÛRVĬCŎS'TĀTE, having curved ribs or large veins.

CÛRVĬDĔN'TĀTE, having curved teeth.

CÛR'VĬFÔRM, curved.

CÛR'VĬNĒRVED, having curved nerves. Applied to the veins (so-called nerves) of endogens.

CÛRVĬSĒ'RĬAL, in curved ranks. Compare RECTISERIAL.

CUSH'ION, a thickened medial portion of the prothallus in ferns; any pulvinus.

CŬSP, a sharp rigid point.

CŬSPED, see CUSPIDATE.

CŬS'PĬDĀTE, furnished with a cusp.

CŬS'PĬS, see CUSP.

CŬT, acutely cleft or parted; having acute incisions deeper than Dentate. Compare CLEFT, INCISED, and LACINIATE.

CŪ'TĬCLE, the outer cell-wall of the epidermis when thickened or otherwise modified.

CŪTĬCŪLÄRĬZĀ'TION, the formation of cuticle.

CŪ'TĬN, see SUBERIN.

CŪ'TĬS, see CUTICLE.

CŪTĬNĬZĀ'TION, the formation of cuticle. Compare SUBERIZATION.

CŬT-TŌOTHED, deeply and sharply toothed.

ÇȲĀ'NĒŬS, pure blue. Compare CÆRULEUS.

ÇȲĂN'ĬC FLOWĔRS, those whose color contains more or less blue. Compare XANTHIC FLOWERS.

ÇȲĂN'ŌPHȲLL, see PHYLLOCYANIN.

ÇȲĂTH'ĬFÔRM, wine-glass shaped. Compare COTYLIFORM and ACETABULIFORM.

ÇȲĂTH'ĬŬM, a corolla-like involucre, as in Euphorbia.

ÇȲ'ĂTHŌID, see CYATHIFORM.

ÇȲ'CLE, a complete turn in a spire or circle.

ÇȲC'LĬC, having the floral organs in distinct whorls. Compare HEMICYCLIC and ACYCLIC.

ÇȲC'LĬCAL, coiled into a full circle.

ÇȲCLŌ'SĬS, see ROTATION.

ÇY̆CLŌSPĔR'MOŬS, having the embryo coiled around the central albumen, as in the Caryophyllaceæ.

ÇY̆LĬNDRĀ'ÇEOŬS, nearly cylindrical.

ÇY̆LĬN'DRĬCAL, circular in transverse outline and tapering but little if at all, as most stems.

ÇY̆'MȦ, see CYME.

ÇY̆M'BÆFÔRM, see NAVICULAR.

ÇY̆M'BĬFÔRM, see NAVICULAR.

ÇȲME, a somewhat flat-topped determinate inflorescence resembling a corymb.

ÇȲME'LĔT, see CYMULE.

ÇȲMĬF'ĔROŬS, producing cymes.

ÇȲMŌ-BŎT'RY̆S, a mixed inflorescence in which the primary inflorescence is botryose while the secondary is cymose, as in the horse-chestnut.

ÇȲ'MOID, having the form of a cyme.

ÇȲ'MŌSE, growing in cymes; cymous; or cymoid.

ÇȲ'MŌSE ĬNFLŌRĔS'ÇENÇE, a cyme or other determinate inflorescence.

ÇȲ'MŌSE ŬM'BĔL, one having the inflorescence centrifugal; a cyme resembling an umbel. Umbels are usually botryose (centripetal) in inflorescence.

ÇȲ'MOŬS, see CYMOSE.

ÇY̆M'ŪLȦ, see CYMULE.

ÇȲ'MŪLE, a little cyme, or division of a compound cyme. Sometimes applied to the verticillasters of Labiatæ.

ÇY̆NĂRRHŌ'DĬŬM, a fruit-like body, like the hip of roses, consisting of several achenia enclosed in a fleshy receptacle.

ÇY̆NĂR'RHŎDŎN, see CYNARRHODIUM.

ÇY̆PĒRĀ'ÇEOŬS, resembling or pertaining to plants of the family Cyperaceæ.

ÇY̆PHĔL'LȦ (pl. **Cyphĕl'læ**), a kind of pit in the under-surface of the thallus of some lichens. Cyphellæ sometimes appear as spots or as elevations. Their nature is unknown.

ÇY̆PHĔL'LĀTE, having Cyphellæ.

ÇY̆P'SĔLȦ, an achenium with an adherent calyx-tube, as in Compositæ. (Obs.)

ÇY̆ST, a closed sac, especially one abnormally produced, or whose nature is not understood.

ÇY̆S'TĬD (pl. **Çys'tids** or **Çys'tĭdēs**), a large sterile club-shaped cell common among the basidia in some agarics.

ÇY̆STĬD'ĬŬM (pl. **Çystĭd'ĭȧ**), see CYSTID.

ÇY̆S'TŌBLȦST (obs.), see NUCLEUS.

ÇY̆S'TŌCÄRP, an old term for the Sporocarp of Florideæ.

ÇY̆S'TŌLĬTH, a cluster of crystals of calcium carbonate in a cell upon a stalk of cellulose. Common in the leaves of the nettle family (Urticaceæ).

ÇY̆TĂS'TĔR, a series of achromatic rays extending from each pole of the nucleus in karyokinesis into the cytoplasm. Rarely seen as yet in plants.

ÇY̆TĔN'CHY̆MȦ, see CELL-SAP.

ÇY̆T'ĬŌDĔRM, the cell-wall in Diatomaceæ.

ÇȲ'TŌBLȦST (Schleiden), see NUCLEUS.

ÇȲTŌDĪĔR'ĒSĬS, cell-division, involving division of the nucleus with the formation of a nuclear-spindle and asters. (Carnoy.) Compare STENOSIS.

ÇYTŌĠĔN'ĔSĬS, cell-formation.

ÇȲTŎL'ŌĠY̆, the science of cells

ÇY̆'TŌPLĂSM, the protoplasm in a cell outside the nucleus. It excludes granules of protein, starch, etc. Compare NUCLEOPLASM.

DÄRWĬN'ĬAN CÛR'VĀTŪRE, the curvature of the growing apex of a root away from any source of irritation placed upon one side near the tip. It is the irritability which causes this curvature that enables growing roots to pass around obstacles in the soil.

DAUGH'TĔR-ÇĔLL, any cell when mentioned in relation to the one from which it was derived. Compare MOTHER-CELL.

DAUGH'TĔR-SPŌRE, a spore produced directly from another, or upon a promycelium.

DAUGH'TĔR-STÄR, one of the groups of chromatic filaments at the poles of a dividing nucleus. The two polar figures together with the connecting spindle-fibres are called a Dyaster.

DĒĂL'BĀTE, appearing as if whitewashed; covered with a very white bloom or powder.

DĔCĂĠY̆N'ĬAN, see DECAGYNOUS.

DĒCĂĠ'Y̆NOŬS, having ten pistils or styles; decagynian.

DĒCĂM'ĔROŬS, having the parts in tens, as ten floral organs in a whorl. Also written 10-merous.

DĔCĂN'DRĬAN, see DECANDROUS.

DĒCĂN'DROŬS, having ten stamens; decandrian.

DĔCÅPĔT'ALOŬS, having ten petals.

DĔCÅPHY̆L'LOŬS, having ten leaves, as a *decaphyllous* perianth.

DĔÇĔMDĔN'TĀTE, having ten teeth or tooth-like processes.

DĔÇĔM'FĬD, ten-cleft.

DĒÇĔMLŎC'ŪLAR, having ten loculi or cells in an ovary.

DĔÇĬD'ŪOŬS, falling at the usual time, or at the close of the season. Applied to leaves which fall in autumn after one season's growth, and to plants which bear such leaves. Also applied to petals which fall immediately after blossoming. Compare CADUCOUS, MARCESCENT, PERSISTENT, and EVERGREEN.

DĔC'LĬNĀTE, bent or curved downward; declining; declined. Applied to stamens it means curved to one side, neither outward nor inward nor erect. Compare RECURVED, REFLEXED, RECLINATE.

DĔCLĬNED', see DECLINATE.

DĒCLĪ'NOŬS, see DECLINATE.

DĒCŎMPOUND', twice compound, as a compound leaf whose parts are compound. See SUPRADECOMPOUND.

DĔCRĒAS'ĬNGLY̆ - PĬN'NĀTE, having the leaflets of a pinnate leaf gradually smaller in size from base to apex.

DĒCŬMBENT, erect at the base, then prostrate, with the end rising.

DĒCŬR'RENT, prolonged and attached below the main point of insertion, as the leaves of thistles; decursive. The stems in such cases are often called Winged. Compare SURCURRENT.

DĒCÛR'SĬVE, see DECURRENT.

DĒCÛR'SĬVELY̆-PĬN'NĀTE, apparently pinnate, but having the segments decurrent along the rachis.

DĒCŬS'SĀTE, said of leaves which grow in pairs successively at right angles to each other. Compare BRACHIATE.

DĒCŬS'SĀTĔD, see DECUSSATE.

DEDOUBLEMENT [Day-dōōbl-i-män], see CHORISIS.

DĒDŪPLĬCĀ'TION, see CHORISIS.

DĔF'ĔRENT, carrying anything downwards.

DĔF'ĬNĬTE, (1) having a constant, fixed, or limited, number; in stamens not exceeding twenty; (2) the same as Determinate, which see.

DĔF'ĬNĬTE GRŌWTH, see DETERMINATE GROWTH.

DĔF'ĬNĬTE ĬNFLŌRĔS'ÇENÇE, see DETERMINATE INFLORESCENCE.

DĔFLĔCT'ĔD, see DEFLEXED.

DĔFLĔXED', bent abruptly to one side or outward at an angle of forty-five degrees or less; deflected. Compare REFRACTED and RECURVED.

DĔFLŌ'RĀTE, past the flowering state, as an anther after it has shed its pollen, or a plant or flower after the petals have fallen.

DĔFŌ'LĬĀTE, having cast its leaves.

DĔFŌLĬĀ'TION, the fall of leaves.

DĔF'ÔRMĀ'TION, an alteration in the usual form of an organ, by accident or otherwise; malformation.

DĔĠĔNĔRĀ'TION, deficiency in size or quality, or appearing in a less developed or lower state, as when scales take the place of leaves, or petals the place of stamens; degradation.

DĔGRȦDĀ'TION, see DEGENERATION.

DĒHĬSÇE', to open for the escape of seeds, spores, etc.

DĒHĬS'ÇENÇE, the opening in a regular manner of certain fruits to discharge their seeds, and of anthers and spore cases to discharge their contents.

DĒHĬS'ÇENT, opening in a regular manner to discharge the contents, as most pods to liberate the seeds. Compare INDEHISCENT.

DĒLĬMĬTĀ'TION, see ABJUNCTION.

DĔLĬQUĔS'ÇENT, dissolving. Said of a tree which branches repeatedly, so that the main stem is lost. Compare EXCURRENT. Also applied to fungi which dissolve at maturity by the absorption of moisture from the air, as some mushrooms.

DĔL'TOĪD, nearly in the shape of an equilateral triangle, or the Greek letter delta. When applied to leaves it implies that the attachment is at the broad end, otherwise the term Obdeltoid is used.

DĒMĔRSED', growing constantly or naturally under water; immersed; submersed.

DĔN'DRĬFŌRM, tree-shaped.

DĔNDRĬT'ĬC, branched like a tree; dendriform.

DĔN'DROĪD, tree-shaped; dendriform; dendritic; arborescent; applied to small plants, such as mosses, which branch like a tree or shrub.

DĔNDRŎL'ŌĠY̆, the natural history of trees.

DĔN'ĬGRĀTE, dark dusky brown.

DĔNSE, crowded together.

DĔN'TĀTE, having broad acute

marginal teeth which are usually directed outward. Compare SERRATE and CRENATE.

DĔN'TĀTĔD, see DENTATE.

DĔN'TĀTE-ÇĬL'ĬĀTE, having the margin dentate and fringed.

DĔN'TĀTE-CRĒ'NĀTE, (1) having part of the margin dentate and the remainder crenate; (2) intermediate between dentate and crenate, i.e., having somewhat rounded teeth.

DĔN'TĀTE-LĂÇĬN'ĬĀTE, having the teeth extended into long lax points.

DĔN'TĀTE-SĔR'RĀTE, (1) intermediate between dentate and serrate; having the teeth directed but slightly forward; (2) having part of the margin dentate and the remainder serrate.

DĔN'TĀTE-SĬN'ŪĀTE, having broad shallow sinuses between the teeth of a dentate margin.

DĔNTĀTŎ-SĔR'RĀTE, see DENTATE-SERRATE.

DĔNTĀTŎ-SĬN'ŪĀTE, see DENTATE-SINUATE.

DĔN'TĬCLE, a small tooth or projecting point.

DĔNTĬC'ŪLĀTE, having small teeth or notches; finely dentate.

DĔNTĬCŪLĀ'TION, (1) the state of being denticulate; toothed; (2) one of the teeth of a denticulate leaf.

DĔN'TĬFÔRM, see DENTOID.

DĔN'TŌID, tooth-shaped.

DĔNŪ'DĀTE, deprived of the usual covering, especially of a coating of down or hair. The term has been applied with little reason to plants whose flowers appear before their leaves.

DĔŌPĔR'CŪLĀTE, applied to moss capsules in which the lid does not separate spontaneously to liberate the spores. Compare DISOPERCULATE.

DĔPAU'PĔRĀTE, said of a plant or part of a plant which is reduced in size from lack of nutriment, or apparently so; starved; stunted; undeveloped.

DĔPĔND', to hang directly downward.

DĔP'LÀNĀTE, flattened vertically and expanded. Both sides may be flattened (Complanate) or only the upper. (Rare.) See EXPLANATE, DEPRESSED, and COMPLANATE.

DĔPRĔSSED', more or less flattened vertically or endwise; oblate. Also applied to any part having a lower position than usual, as a leaf whose disk is lower than the margin, as in some water lilies.

DĔPRĔSSED'-GLŌ'BŌSE, globular with the poles slightly flattened; oblately spheroidal.

DĔRĬV'ÀTĬVE-HȲ'BRĬD, a hybrid of which one or both the parents is a hybrid; secondary hybrid.

DĔRMĂT'ŌGEN, the primordial epidermis, i.e., the epidermis in an organ when first formed and still capable of cell-division.

DĔRMĂT'ŌPHȲTE, any fungus parasitic upon the skin of man or other animals, as *Trichophyton tonsurans*, the fungus which causes ringworm.

DĔSÇĔND'ĬNG, directed downward in any degree.

DĔSÇEND'ĬNG ĂX'ĬS, root.

DĔSÇĔND'ĬNG MĔTĂMÔR'PHŌSĬS, see RETROGRESSIVE METAMORPHOSIS.

DĔSÇĔND'ĬNG SĂP, a term formerly applied to cambium

when in a soft mucilaginous condition, from the supposition that the sap passed in a current upward in the wood to the leaves, then downward between the wood and bark. See PROPER JUICE.

DĔSÇRĬP'TĬVE BŌT'ÅNY̆, see PHYTOGRAPHY.

DĔSTRŬC'TĬVE MĔTĂB'ŌLĬṢM, the chemical changes which take place during the waste of tissues. Compare ASSIMILATION.

DĔSTRŬC'TĬVE PĂR'ÅSĪTE, one which causes the death of the tissues of the host upon which it feeds.

DĔTĔR'MĬNĀTE GRŌWTH, when the season's growth ends in a well-formed bud.

DĔTĔR'MĬNĀTE ĬNFLŌRĔS'ÇENÇE, one in which the flowering begins with a terminal bud so that it puts a limit to the elongation of the stem; centrifugal or definite inflorescence.

DĔTĔRMĬNĀ'TION, the discovery of the botanical name of a plant, or of its position in a system if unnamed; identification. It implies some knowledge of its characters acquired by a more or less complete analysis, and is frequently completed by the aid of a systematic key.

DEŬ'TŌPLĂṢM, the portion of the cell contents aside from the protoplasm proper, consisting of cell-sap in which may be granules or other substances; paraplasm. Compare METAPLASM.

DĔX'TRŌRSE, twining upward from left to right, with the sun or hands of a watch, as the hop. Used in the opposite sense, however, by most English and American botanists. Compare SINISTROSE.

DĪĂCHÆ'NĬŬM, see MERICARP.

DĪĂCH'Y̆MÅ (obs.), see MESOPHYLLUM.

DĪĂDĔL'PHĬAN, see DIADELPHOUS.

DĪĂDĔL'PHOŬS, having filaments united by their edges in two sets (one of which may be only a single stamen), as in the pea.

DĪĂĠĔŎT'RŌPĬṢM, a kind of geotropic irritability in certain organs, as rootstocks, which causes them to assume a horizontal position; transverse geotropism.

DĪĂGNŌ'SĬS (pl. **Dīăgnō'sēs**), a brief characteristic description of a plant or species.

DĪĂGNŌS'TĬC CHĂR'ĂCTĔR, see ESSENTIAL CHARACTER.

DĪĂG'ŌNAL PLĀNES, the two planes which bisect the right angles between the median and lateral planes. See MEDIAN PLANE and LATERAL PLANE

DĪĂG'ŌNAL PŌṢĬ'TION, a position intermediate between the median and the lateral plane and bisected by the diagonal plane.

DĪ'ÅGRĂM, FLŌ'RAL, see FLORAL DIAGRAM.

DĪÅHĔLĬŎT'RŌPĬṢM, the tendency of organs to place their surfaces at right angles to the sun's rays, as most leaves; transverse heliotropism.

DĪĂLY̆CÄR'POŬS, see APOCARPOUS.

DĪĂLY̆PĔT'ALOŬS, see POLYPETALOUS.

DĪĂLY̆PHY̆L'LOŬS, see POLYSEPALOUS.

DĪĂLY̆SĔP'ALOŬS, see POLYSEPALOUS.

DĪĂL'Y̆SĬS, the separation of organs usually joined; adesmy.

When applied to parts of a flower it refers to the separation of organs of the same whorl from each other, as when a normally gamopetalous corolla has its petals distinct. Compare SOLUTION.

DĪAMESŎG'ÅMOŬS, fertilized by the aid of some external agent, as wind, water, or insects.

DĪĂN'DRĬAN, see DIANDROUS.

DĪĂN'DROŬS, having two stamens.

DĪĂPH'ÅNOŬS, transparent, or nearly so.

DĪ'ÅPHRĂGM, any dividing membrane or partition; septum. The term is usually applied to the septum at the node of a hollow-stemmed grass.

DĪĂP'HȲSĬS, an old term for the proliferation of a flower.

DĪ'ÅSTÅSE, a ferment generated in germinating seeds, etc., which assists in the transformation of starch into sugar.

DĪĂS'TĔR, see DYASTER.

DĪĂT'ŌMĬNE, see PHYCOXANTHINE.

DĪCÄR'PĔLLĀRȲ, of two carpels; digynous.

DĪCHĀ'SĬŬM (pl. **Dīchā'sĭå**), a false dichotomy in which two lateral shoots of nearly equal strength arise beneath the apex which bears a flower; biparous, dichotomous, or forked cyme. Compare COMPOUND DICHASIUM.

DĪCHĂS'TĬC, spontaneously dividing.

DĪCHLĂMȲD'ĔOŬS, having both calyx and corolla.

DĪCHŌG'ÅMOŬS, having flowers in which the stamens and pistils mature at different times; either protandrous or protogynous. Compare SYNACMIC.

DĪCHŌT'ŌMAL FLOWER, one seated in the fork of a dichasium.

DĪCHŌT'ŌMĪZE, to fork.

DĪCHŌT'ŌMOŬS, forked; furcate; bifurcate.

DĪCHŌT'ŌMȲ, forking into two branches of the same nature and usually of about the same size. True dichotomy is caused by the cessation of the previous increase in length of a member at the apex, and its continuance in two diverging directions from two newly constituted apices. Compare HELICOID and SCORPIOID DICHOTOMY.

DĪCLĔ'SĬŬM, a name formerly applied to an achenium having an adherent calyx, as in Mirabilis.

DĪC'LĬNŌUS, having the stamens and pistils in separate blossoms —either monœcious or diœcious; separated; digamous; unisexual.

DĪCŌC'COŬS, having an ovary consisting of two closed carpels or cocci with one seed in each, as in the Umbelliferæ.

DĪÇŒ'LOŬS, having two cavities.

DĪCŌTȲLĔD'ŌNOŬS, having two cotyledons.

DĬD'ȲMOŬS, in pairs.

DĬDȲNĀ'MĬAN, see DIDYNAMOUS.

DĬDȲN'ÅMOŬS, having two long and two short stamens. Compare TETRADYNAMOUS.

DĪĒ'ÇIOŬS, see DIŒCIOUS.

DĬF'FLŬENT, readily dissolving. Compare DELIQUESCENT.

DĬFFŌRMED', of unusual shape.

DĬFFRĂCT', said of a lichen thallus which is broken into areolæ separated by clefts or chinks.

DIFFUSE', spreading widely, loosely, and irregularly. Compare EFFUSE.

DIFFU'SION, the spontaneous mixture of one liquid with another, or of any dissolved or finely divided substance or any liquid through a liquid or solid. Compare OSMOSE.

DIG'AMOUS, see DICLINOUS.

DIG'ENOUS, containing both sexes or produced sexually. (Rare.)

DIG'ITATE, having several parts radiating from one point, or nearly so, somewhat like the fingers on one's hand. Applied in grasses where several spikes radiate from nearly the same point, as in crab-grass (Sanguinaria). In leaves it is the same as Palmately-compound. Compare PEDATE.

DIG'ITATE PIN'NATE, having a digitate leaf with pinnate leaflets.

DIG'ITATE-VEINED, see PALMATELY-VEINED.

DIG'ITIFORM, see DIGITATE.

DIG'ITINERVED, see PALMATELY-VEINED.

DIG'ONOUS, two-angled, as the stems of some cacti.

DIGYN'IAN, see DIGYNOUS.

DIG'YNOUS, having two pistils, styles, or stigmas in a flower. Generally the same as Dicarpellary.

DILAMINA'TION, see CHORISIS.

DILAT'ED, expanded or widened.

DIM'EROUS, having the parts in twos, as two sepals, two petals, two stamens, and two pistils in a flower.

DIMID'IATE, halved, or appearing as if one side or one half were wanting, as a leaf which is developed on one side of the midrib only. Compare OBLIQUE. Applied also to the calyptra of mosses when divided along one side. The term is also applied to bodies which are actually halved or divided into two parts, and to those in which one side differs from the other in function.

DIMOR'PHOUS, existing under two forms, as two forms of leaves, sterile and fertile, on one plant in ferns. Applied especially to heterogonous plants having two forms of flowers on separate individuals, one with long stamens and short styles and another with short stamens and long styles. Compare TRIMORPHOUS.

DIŒ'CIAN, see DIŒCIOUS.

DIŒ'CIOUS, having stamens and pistils in separate flowers upon different individuals.

DIŒ'CIOUSLY POLYG'AMOUS, when some of the flowers in diœcious plants are perfect. Each set of plants may have perfect as well as unisexual flowers, or one set may have all perfect flowers and the other all staminate or all pistillate; polygamo-diœcious.

DI'OSMOSE, see OSMOSE.

DIPET'ALOUS, having two petals in a flower.

DIPHYL'LOUS, two-leaved. Having two leaves, leaflets, or leaf-like parts. Sometimes used for Disepalous.

DIPLAN'ETISM, the property of being twice active with an intervening period of rest. It occurs in the zoöspores of certain genera of Saprolegnieæ in which the zoöspores escape without cilia from the sporangium and come to rest in a cluster each forming a cell-wall. After some hours of rest the protoplasm of each spore escapes from its cell-wall, acquires cilia, and enters upon a period of active movement.

DĬP'LŌĔ, see MESOPHYLL.

DĬPLŌĠĔN'ĔSĬS, the duplication of parts normally single.

DĬPLŌPĔRĬS'TŌMOŬS, said of the capsule in mosses when the peristome consists of a double row of teeth. Compare APLOPERISTOMOUS.

DĬPLŌSTĔM'ŌNOŬS, having the stamens in two whorls, those of each whorl equal in number to the petals, which are in one whorl. Compare ISOSTEMONOUS.

DĬPLŌTĒ'ĠĬŬM (pl. **Dĭplōtē'gĭà**), an inferior capsule, as in Campanula.

DĬP'TĔROŬS, two-winged.

DĬREC'TION-ÇĔLLS, see POLAR CELLS.

DĬRĔC'TION-CÔR'PŬSÇLES, see POLAR CELLS.

DĬRĔCT' MĔTĂMÔR'PHŌSĬS, see PROGRESSIVE METAMORPHOSIS.

DĬRĔCT' SŬPĔRPŌȘĬ'TION, the situation of accessory buds in an axil above the leading bud or one first formed, as is usually the case in superposition. Compare INVERTED SUPERPOSITION.

DĬRĔMP'TION, see DISPLACEMENT.

DĬSÄRTĬC'ŪLĀTE, to separate at an articulation or joint, as most leaves in autumn.

DĬSC, see DISK.

DĬS'ÇĬFÔRM, disk-shaped; flat and circular.

DĬSÇĬG'ĔROŬS, disk-bearing, as the woody tissue of conifers.

DĬS'CŌCÄRP, an ascocarp in which the hymenium lies exposed while the asci are maturing, as in Peziza. The term has also been applied to fruits like the rose and fig, in which the receptacle is expanded and forms the main part of the fruit, enclosing the seeds or achenia.

DĬSCOID, disk-shaped; flat and circular; belonging to the disk, as the central florets in a head of Compositæ.

DĬS'COID FLOWĔR, a head of flowers in Compositæ in which each flower is tubular.

DĬS'COID MÄRK'ĬNG, see BORDERED PIT.

DĬSCŌL'ŌR, of more than one color; variegated; especially having the two surfaces of a leaf differing in color, as in the begonias. Compare UNICOLOR and CONCOLOR.

DĬS'COŬS, disk-shaped.

DĬSCRĒTE', distinct; separate. Compare CONCRETE.

DĬS'CŬS, see DISK.

DĬSĔP'ALOŬS, of two sepals.

DĬSK, any flat, circular area: the central part of such an area or of any flat body as opposed to the border; disc. Especially (1) the central part of a head of flowers in Compositæ, generally bearing tubular florets only; (2) the portion of the receptacle of a flower between the stamens and pistil, often more or less dilated and serving as a nectary; (3) a circular bordered pit, as in Coniferæ; (4) the circular adhesive base (retinaculum) of a pollinium.

DĬSK FLŌ'RĔT, see TUBULAR FLORET.

DĬSK' FLOWĔR, see TUBULAR FLORET.

DĬSK'-SHĀPED, flat and more or less circular; discoid.

DĬSLŌCĀ'TION, see DISPLACEMENT.

DĬSŌPĔR'CŬLĀTE, having lost the operculum or lid. Compare DEOPERCULATE.

DĬSPLĀÇE'MENT, the situation of an organ out of its normal position; dislocation; diremption.

DĬSSĔCT'ĔD, cut deeply into many lobes or divisions. Compare LACINIATE and DIVIDED.

DĬSSĔMĬNĀ'TION, the natural dispersion of seeds.

DĬSSĔP'ĬMENT, one of the double walls separating the cells of a syncarpous ovary; septum. Compare FALSE DISSEPIMENT.

DĬSSĬL'ĬENT, dehiscing with elastic violence, as the pod of Impatience.

DĬSSŌÇĬĀ'TION, separation.

DĬS'TAL, pertaining to the apex or outer extremity. Compare PROXIMAL.

DĬS'TANT, having larger intervening spaces than usual. Opposed to Dense or Approximate.

DĬS'TĬCHOŬS, two-ranked, as the leaves of grasses. Applied also to leaves arranged like those of the fir, which are turned in two directions, though not inserted oppositely in the same plane. In this latter sense Bifarious is somewhat the better term.

DĬSTĬNCT', when parts of the same kind are unconnected; opposed to Coherent. Compare FREE.

DĬSTRĂC'TĬLE, widely separated. Applied mainly to anthers in which the connective is developed so as to keep the lobes wide apart, as in Salvia.

DĪTRĪCHŎT'ŎMOŬS, dividing into two or three branches.

DĪŬR'NAL, said of flowers which open in the day and close at night.

DĪŬR'NAL SLĒEP, see PARAHELIOTROPISM.

DĬVĂR'ĬCĀTE, diverging at a wide angle.

DĬVĔR'ĠENÇE, see ANGLE OF DIVERGENCE.

DĬVĔRSĬFLŌ'ROŬS, having flowers of two or more forms.

DĬVĪD'ĔD, having incisions extending to the midrib. Compare CLEFT and PARTED.

DŌDĔCĂĠ'ȲNOŬS, having twelve pistils.

DŌDĔCĂM'ĔROŬS, having the floral organs in twelves; 12-merous.

DŌDĔCĂN'DROŬS, having twelve stamens.

DŌDĔCĂPĔT'ĂLOŬS, having twelve petals.

DŌLĂB'RĬFŌRM, having the form of an axe or hatchet, as the leaves of *Mesembryanthemum dolabriforme.*

DŌMĔS'TĬCĀTĔD, introduced and found to thrive and reproduce itself under cultivation. It does not necessarily imply any change of character.

DŌR'MANT BŬD, one which is poorly developed and which under ordinary circumstances will not grow into a branch. Often the first-formed buds on a season's growth are of this character; latent bud.

DŌR'MANT STĀTE, the condition of a living plant during winter, or other definite period of cessation from active growth; latent period.

DŌR'SAL, pertaining to the back, or situated upon the back. The dorsal surface of a leaf or other foliar organ is the anterior, outer, or normally lower surface. Some botanists, however, apply the term dorsal to the upper surface. Compare VENTRAL.

DŌR'SAL SŬ'TŬRE, one situated at the midrib of the carpellary leaf.

DÔRSĬCŬM'BENT, see SUPINE.

DÔRSĬF'ĔROŬS, bearing the fruit upon the back, as the spores of many ferns; dorsiparous.

DŌRSĬP'ĂROŬS, see DORSIFEROUS.

DŌRSĬVĔN'TRAL, having a distinct front and back, as leaves.

DŌR'SŬM, the back of any organ: the outer or lower surface of leaves or parts of a flower. See also DORSAL.

DŌR'TȲ, (Hort.), delicate; difficult to cultivate. (Rare.)

DŎT'TĔD, see PUNCTATE

DŎT'TĔD DŬCT, see PITTED VESSEL.

DOŬB'LE, having more than one whorl of petals. A flower is completely double when all the essential organs are replaced by petals.

DOŬB'LȲ-CŎM'POUND, twice compound, as bipinnate or bipalmate.

DOŬB'LȲ-CRĒ'NĀTE, having the crenatures or teeth of a crenate leaf again crenate.

DOŬB'LȲ-DĔN'TĀTE, having the teeth of a dentate leaf dentate. Compare BIDENTATE.

DOŬB'LȲ-PĬN'NĀTE, see BIPINNATE.

DOŬB'LȲ-SĔR'RĀTE, having small serratures upon the large ones, as in the elm. Compare BISERRATE.

DOŬB'LȲ-TĔR'NĀTE, see BITERNATE.

DOŬB'LȲ-TŌŌTHED, having the teeth themselves toothed, as in Doubly-dentate.

DOWN, soft short pubescence.

DOWN'Ȳ, having a dense covering of short weak hairs.

DRAWN, elongated by absence of light, as plants which are crowded together.

DRĔP'ÅNĬFŌRM, see FALCATE.

DRŌŌP'ĬNG, inclining downward more than cernuous and less than pendent or pendulous.

DRṲPĀ'ÇEOŬS, like, or pertaining to, a drupe; producing drupes, as *drupaceous* trees.

DRṲPE, a stone-fruit, as the peach, almond, and cherry. It consists of one carpel, with usually a single seed surrounded by a thickened bony endocarp called the pit or stone.

DRṲ'PEL, a very small drupe, as the so-called seeds of the blackberry.

DRṲPE'LĔT, see DRUPEL.

DŪ'BĬOŬS, doubtful.

DŬCT, see VESSEL.

DŬL'CĬS, devoid of acidity: sweet.

DŪMŌSE', having the form of a bush or low compact shrub.

DŪ'MŬS, see BUSH.

DŪ'PLĬCĀTE, doubled or folded.

DŪ'PLĬCĀTE-CRĒ'NĀTE, see DOUBLY-CRENATE.

DŪ'PLĬCĀTE-DĔN'TĀTE, see DOUBLY-DENTATE.

DŪ'PLĬCĀTE-SĔR'RĀTE, see DOUBLY-SERRATE.

DŪRĀ'MĔN, heart-wood; the central portion of the trunk of most exogens, consisting of wood of darker color and denser texture than the outer newer layers, and possessing the characteristic color of the species.

DWARF, habitually attaining much less than the ordinary size of related species or varieties; nanus. Compare HUMILIS and DEPAUPERATE.

DWARF MALES, very small individuals in Œdogonieæ, originating from special swarmspores called androspores, and producing only antherozoids.

DYASTER, a stage of karyokinesis succeeding the monaster or mother-star stage and ending with the formation of the daughter-skeins. During this stage the chromatic filaments are grouped about the poles, the two groups (asters) being more or less united by the spindle fibres or conjunctive threads. See DAUGHTER-STAR.

Ĕ- or ĔX-, a prefix meaning destitute of, outside of, or away from.

ĒAR, a prominent lobe, as those at the base of the leaf in sorrel (*Rumex acetocella*).

ĒARED, see AURICULATE.

ĒBĔ'NĔOŬS, black like ebony.

ĒBRĂC'TĒĀTE, without bracts.

ĒBŬR'NĔOŬS, ivory-white.

ĒCĂL'CĀRĀTE, without a spur.

ĒCAU'DĀTE, without a tail or tail-like appendage.

ĔCBLĂSTĔ'SĬS, the production of buds within a flower in consequence of lateral prolification.

ĔCÇĔN'TRĬC, out of the centre or axis; not having the same centre; when the centre or axis of growth does not coincide with that of the mass, as in most starch-grains, or in trees which develop more rapidly on one side than on the other.

ĔCH'ĬNĀTE, spiny or prickly.

ĔCHĬN'ŬLĀTE, diminutive of Echinate; having small prickles.

ĔÇĬD'ĬŬM, see ÆCIDIUM.

ĒCŌNŌM'ĬC BŎT'ĂNY̆, the classification of useful and injurious plants, and the study of all botanical questions having a practical bearing. See AGRICULTURAL BOTANY.

ĒCŌS'TĀTE, without a rib.

ĔCTŌĠĔ'NĬC, capable of living outside of an animal body. Said of certain disease-producing organisms, as the bacillus of anthrax.

ĔC'TŌPLĂSM, a hyaline layer of protoplasm free from granules next to the cell-wall; hyaloplasm.

ĔC'TŌSPŌRE, see BASIDIOSPORE.

ĔCTŌSPŌ'ROŬS, see EXOSPOROUS.

ĔCTŌTHĒ'CAL, gymnocarpous, as applied to Ascomycetes. (Rare.)

ĒDĔN'TĀTE, without teeth, as an entire leaf.

ĔDĠED, see MARGINATE.

ĔFFĒTE', exhausted; no longer productive or fruitful; past the bearing age.

ĔFFLŌRĔS'ÇENÇE, the time or act of flowering.

ĔFFŪSE', spreading loosely, especially on one side, as the panicle of *Juncus effusus*. Compare DIFFUSE and SECUND.

ĔGG-ĂPPĀRĀ'TŬS, a group of three nucleated bodies at the upper end of the embryo-sac, which together with the upper polar nucleus correspond with the antipodal cells. The central deeper cell of the egg-apparatus becomes the oösphere, the others form the elongated synergidæ; germinal apparatus.

ĔGG-ÇĔLL, see OÖSPHERE.

ĒFŬL'CRĀTE, said of buds from below which the leaf has fallen. (Rare.)

ĒGLĂN'DŪLŌSE, without glands.

Ē'GRĔT, see AIGRET.

EĪS'ŌDAL, anterior. Applied to the outer or anterior part of the opening of a stoma. Compare OPISTHODAL.

ĒJĔC'TION, throwing out with force, as spores from a sporangium or seeds from a pod. Compare ABJECTION.

ĒLĂBŌRĀ'TION, a term applied to the changes which take place in plant-food after it is absorbed, to prepare it for the use of the plant. Compare ASSIMILATION and METASTASIS.

ĔL'ĀTĔR, a term for various elastic, usually spiral, bodies which serve for the dispersion of spores, as those in the capsules of mosses and liverworts and those attached to the spores of Equisetum.

ĔLĀTĒ'RĬŬM, a term sometimes applied to fruits which discharge their seeds by means of elastic carpels, as in certain Euphorbias; regma.

ĔLEŪTHĔRŌPĔT'ĂLOŬS, see APOPETALOUS.

ĔLEŪTHĔRŌPHȲL'LOŬS, see APOPHYLLOUS.

ĔLEŪTHĔRŌSĔP'ĂLOŬS, see APOSEPALOUS.

ĔLLĬPSOĪD'AL, nearly elliptical in outline.

ĔLLĬP'TĬCAL, oblong, and rounded at the ends; longer than oval.

ĔLLĬP'TĬC-LĂN'ÇĔŌLĀTE, intermediate between elliptical and lanceolate, but approaching the latter.

ĒLŎC'ŪLAR, see UNILOCULAR.

ĒLŎN'GĀTĔD, exceeding the usual or average length.

ĔLȲT'RĬFŌRM, resembling the wing-cover of a beetle; elytroid.

ĒMĂR'ÇĬD, flaccid; wilted. (Obs.)

ĒMĂR'ĠĬNĀTE, notched at the end. Sometimes applied to organs which are notched at other places, as to the gills of mushrooms when notched or cut out before reaching the stem.

ĔMBŌSSED', see UMBONATE.

ĔMBRĀÇ'ĬNG, clasping by a broad surface. Compare AMPLECTANT and SHEATHING.

ĔM'BRȲŌ, the rudimentary plant within the seed.

ĔM'BRȲŌ-BŬD, a rudimentary bud, especially if adventitious.

ĔM'BRȲŌ-ÇĔLL, see OÖSPHERE.

ĔMBRȲŌĠĔN'ĬC, pertaining to the development of an embryo.

ĔMBRȲŎĠ'ENȲ, embryo-formation.

ĔMBRȲŎL'ŌĠȲ, the study of the embryo and its development.

ĔM'BRȲŎNAL ÇĔLL, see OÖSPHERE.

ĔM'BRȲŌNAL VĔS'ĬCLE, see OÖSPHERE.

ĔM'BRȲŌNĀTE, having an embryo.

ĔMBRȲŎN'ĬC, in an early undeveloped condition; rudimentary.

ĔM'BRȲŌ NŎD'ŪLE, a term applied to small knots, frequently about the size of a pea, found beneath the bark in certain trees, and sometimes containing one or more rudimentary buds.

ĔMBRY̆ŌN'ĬC SĂC, see EMBRYO-SAC.

ĔMBRY̆ŌN'ĬC VĔS'ĬCLE, see OÖSPHERE.

ĔM'BRY̆Ō-SĂC, a large cell in the nucleus of the ovule within which the germinal vesicles or oöspheres (one or more) are produced, and which finally contains the embryo. Compare CENTRAL CELL.

ĔMBRY̆ŌTĒ'ĠĬĂ (pl.), see EMBRYOTEGIUM.

ĔMBRY̆ŌTĒ'ĠĬŬM (pl. Ĕmbry̆ō-tē'ġĭă), a small cap covering the micropyle in certain seeds, as asparagus, and detached by the radicle in germination.

ĒMĔR'ĠENÇĬEȘ, a term applied to outgrowths of various kinds derived from the fundamental tissue below the epidermis, and covered by the latter, as the prickles of the rose.

ĒMĔR'ĠENT, protruding through or elevated above surrounding parts.

ĒMĔRSED', raised out of water.

ĔMPĀLE'MENT, an old term for calyx.

ĔMP'TY̆ GLŪMEȘ, one, two, or more bracts or scales subtending a spikelet in grasses, and enclosing one or more flowers; outer glumes. Formerly called merely glumes. Compare FLOWERING GLUME.

ĔNĂNTĬŌBLĂS'TĬC, a term sometimes applied to the embryo of orthotropous seeds. Compare HOMOBLASTIC.

ĒNĀ'TION, having outgrowths or excrescences the result of excessive development, as scales upon petals. Compare CHORISIS.

ĔNCHY̆LĒ'MĂ, the unorganized proteids in living cells, as aleurone grains. (Hanstein.)

ĔNÇY̆S'TĔD, enclosed in a cyst or sac. Applied, for example, to a stage of growth in Protococcaceæ in which the individual exists as a free cell with a cell-wall, but destitute of cilia.

ĔNDĔCĂĠ'Y̆NOŬS, having eleven pistils or styles.

ĔNDĔCĂN'DROŬS, having eleven stamens.

ĔNDĔCĂPHY̆L'LOŬS, said of a leaf containing eleven leaflets.

ĔNDĔM'ĬC, occurring in the one limited locality or region only. Compare SPORADIC.

ĔNDŌBĂSĬD'ĬŬM, an enclosed basidium; as in Gasteromycetes.

ĔN'DŌCÄRP, the inner layer of a pericarp, particularly if developed in a special manner, as the pit of a peach or core of an apple.

ĔN'DŌCHRŌME, coloring matter in cells, or colored cell-contents aside from chlorophyll—used mainly in algæ. Often applied to the entire cell-contents of algæ, and sometimes to colored cell-contents in other plants, but less used now than formerly.

ĔNDŌDĔR'MĬS, a layer, of one or more cells in thickness, which forms the inner boundary of the cortex and surrounds the fibrovascular cylinder.

ĔNDŎĠ'ĔNOŬS, produced within another body. Applied also to the stems of monocotyledons and their manner of growth, which was formerly supposed to take place chiefly at or near the centre.

ĔNDŎĠ'ĔNOŬS ÇĔLL-FŌRMA'-TION, see FREE CELL-FORMATION.

ĔNDŌGŌNĬD'ĬŬM, a gonidium formed within a receptacle, as in the sporangium of Mucorini.

ĔNDŌPĔRĬD'ĬŬM, the inner peridium when there are more than one, as in Geaster.

ĔNDŌPHLŒ'ŬM, inner bark; liber. See BAST.

ĔNDŌPHȲL'LOŬS, enclosed in a leaf or sheath, as the young leaves of monocotyledons.

ĔN'DŌPHȲTAL, growing within other plants; entophytal.

ĔN'DŌPHȲTE, a plant which grows within another, either parasitic upon it or not; entophyte.

ĔN'DŌPLĂṢM, the internal granular portion of the protoplasm; when distinguished from an outer layer free from granules called Ectoplasm.

ĔNDŌPLEŬ'RÀ, see TEGMEN.

ĔNDŎP'TĬLE, an old term applied to the plumule of endogens.

ĔNDŌRHĪ'ZÀ (pl. **Ĕndōrhī'zæ**), an old term for endogen.

ĔNDŌRHĪ'ZAL, said of an embryo in which the radicle is sheathed by the cotyledons wrapped around it. Applied also to the method of germination in endogens.

ĔNDŌRHĪ'ZOŬS, see ENDORHIZAL.

ĔN'DŌSMŌSE, an inward current established between fluids of different densities when separated by an animal or vegetable membrane; endosmosis. The absorption of moisture by roots is due to endosmose. See OSMOSE.

ĔNDŌSMŌ'SĬS, see ENDOSMOSE.

ĔN'DŌSPĔRM, the albumen of the seed, especially when formed in the embryo-sac. Compare PERISPERM.

ĔN'DŌSPŌRE, the inner coat of a spore.

ĔN'DŌSTŌME, the orifice of the inner coat of the ovule, i.e., the inner portion of the foramen. Compare EXOSTOME.

ĔNDŌTHĒ'ÇĬŬM, the lining (tapetum) of an anther cell, consisting of one or more layers within the exothecium.

ĔNDŌZŌ'ĬC, living inside an animal; entozoic.

ĔNNĔĂǴ'ȲNOŬS, having nine pistils or styles.

ĔNNĔĂN'DRĬAN, see ENNEANDROUS.

ĔNNĔĂN'DROŬS, having nine stamens.

ĔNNĔĂPĔT'ALOŬS, having nine petals.

ĔNNĔĂSĔP'ALOŬS, having nine sepals.

ĔNNĔĂSPĔR'MOŬS, nine-seeded.

ĔNŌ'DAL, without nodes.

ĔN'SĀTE, see ENSIFORM.

ĔN'SĬFŌRM, sword-shaped; straight, or nearly so, two-edged, and tapering very gradually from base to apex, as the leaves of iris; gladiate.

ĔNTĪRE', having margins destitute of teeth or notches.

ĔNTŌMŌǴ'ĔNOŬS, growing upon insects, as certain fungi; entomophytous.

ĔNTŌMŌPH'ĬLOŬS, adapted to pollination by insects.

ĔNTŌMŌPH'ȲTOŬS, see ENTOMOGENOUS.

ĔNTŌPĂR'ÀSĪTE, a parasite living entirely within its host.

ĔN'TŌPHȲTAL, see ENDOPHYTAL.

ĔN'TŌPHȲTE, see ENDOPHYTE.

ĔNTŌZŌ'ĬC, see ENDOZOIC.

ĔN′VĔLŌPE-ĂPPÀRĀ′TŬS, all of the ascocarp except the ascus-apparatus.

ĔPĂN′ŌDY̆, a general term for reversion from an irregular to a regular condition. In flowers it is termed "regular peloria."

ĔPĂN′THOŬS, growing upon flowers, as certain fungi.

ĔP′ĔN, see EPENCHYMA.

ĔPĔN′CHY̆MÀ, cambium, and all tissue arising from it. (Nägeli.) Compare FROTENCHYMA.

ĔPHĔM′ĔRAL, lasting but a day or a very short time, as the corolla of purslane.

ĔP′ĬBLÀST, a term applied to a small scale-like appendage in front of the embryo and opposite the scutellum in the seeds of rice and many other grasses. (F. L. Scribner.)

ĔPĬBLĔ′MÀ, a collective name for the epidermal cells of a young organ, especially a root, including the root-hairs. (Obs.) See EPIDERMIS and DERMATOGEN.

ĔPĬCĀ′LY̆X, an involucre resembling an exterior calyx, as in mallow.

ĔP′ĬCÀRP, the outer layer of a pericarp.

ĔP′ĬCHĬL, see EPICHILIUM.

ĔP′ĬCHĬLE, see EPICHILIUM.

ĔPĬCHĬL′ĬŬM, the upper or distal portion of the labellum of an orchid, when especially different from the lower or basal portion. Compare HYPOCHILIUM.

ĔPĬCLĬ′NAL, seated upon the receptacle.

ĔPĬCŌR′MĬC, applied to side branches which develop on the body of a forest tree from which surrounding trees have been removed.

ĔPĬCŌRŌL′LĬNE, upon the corolla.

ĔPĬCŌT′Y̆L, the portion of a young stem between the cotyledons and the lowest true leaves. Compare CAULICLE.

ĔP′ĬDĔRM, see EPIDERMIS.

ĔPĬDĔR′MAL, pertaining to the epidermis.

ĔPĬDĔR′MĬS, the external layer of cells in a plant. Compare CUTICLE and DERMATOGEN.

ĔPĬDĔRMOĪ′DAL LĀY′ĔR, a term sometimes applied to an outer layer of cortical cells bordering on the epidermis.

ĔPĬGÆ′AN, upon or above ground; growing on land in distinction from water; growing close to the earth, as some leaves; rising above ground instead of remaining beneath, as the cotyledons of beans; epigæous.

ĔPĬGĒ′AL, see EPIGÆAN.

ĔPĬGĔN′ĔSĬS, (1) originating upon another body; (2) the theory that the embryo is the joint product of both sexes, as distinguished either from the doctrine that the male parent furnishes the germ and the female simply the *nidus* or resting-place in which it is nourished, or from the theory that the female furnishes the germ which is merely quickened by the influence of the male. Seldom used in botany.

ĔPĬG′ĔNOŬS, growing upon the surface or on the upper surface, as a fungus on its host. Compare HYPOGENOUS and EDOGENOUS.

ĔPĬGĒ′OŬS, see EPIGÆAN.

ĔP′ĬGŌNE, see EPIGONIUM.

ĔPĬGŌ'NĬŬM, the archegonium in mosses after the capsule has developed. It is frequently ruptured, a part being carried up to form the calyptra, and a part remaining at the base of the sporangium or its stalk, as the vaginule. Also an homologous sac enclosing the young sporangium in Hepaticæ.

ĔPĬG'ȲNOŬS, growing upon the ovary.

ĔPĬNÀS'TĬC, denoting curvature resulting from growth on the upper side of an organ making it curve downward. Compare HYPONASTIC.

ĔP'ĬNĂSTȲ, that state of a growing dorsi-ventral organ in which the upper surface grows more rapidly than the lower.

ĔPĬPĔT'ALOŬS, upon the corolla.

ĔPĬPHLŒ'ŌDAL, upon the epidermis or outer surface. Compare HYPOPHLŒODAL.

ĔPĬPHLŒ'ŬM, see CORKY LAYER.

ĔP'ĬPHRĂGM, a membrane closing the mouth of the capsule in mosses; also a delicate membrane closing the cup-like sporophore in Nidularia.

ĔPĬPHȲL'LOŬS, growing or inserted upon a leaf, or upon the upper side. Compare HYPOPHYLLOUS.

ĔP'ĬPHȲTAL, see EPIPHYTIC.

ĔP'ĬPHȲTE, a plant growing upon another but not nourished by it; air-plant. Compare PARASITE.

ĔPĬPHȲT'ĬC, growing upon another plant but not nourished by it; epiphytal; pseudo-parasitic.

ĔPĬPHȲTŌT'ĬC, a term applied by Erwin F. Smith to any wide-spreading disease among plants, corresponding to an epidemic among men or an epizoötic among animals.

ĔP'ĬPLĂSM, protoplasm which remains in an ascus or other unicellular sporangium after the formation of the spores.

ĔPĬP'TĔROŬS, winged at the apex.

ĔPĬRRHĒŎL'ŌĠȲ, the portion of physiological botany which treats of the effects of external agents on living plants. (Rare.)

ĔPĬRRHĪ'ZOŬS, growing upon roots.

ĔP'ĬSPĔRM, the seed-coat, especially the outer coat or testa.

ĔPĬSPŌRĂN'ĠĬŬM, an old term for indusium.

ĔP'ĬSPŌRE, the outer coat of a spore; exospore.

ĔPĬS'TRŌPHĔ, said of chlorophyl-bodies when they take their position along the outer wall of the cell, usually when the light is of medium intensity. Compare APOSTROPHE and SYSTROPHE.

ĔPĬTHĂL'LĬNE, growing upon the thallus.

ĔPĬTHĔ'LĬŬM, any distinct layer of one or more cells in thickness bounding an internal cavity. Formerly applied also to delicate epidermis of rootlets and other external parts.

ĔPĬX'ȲLOŬS, growing upon wood, as many fungi.

ĔPĬZŌ'ĬC, growing upon living animals, either parasitic or not.

Ē'QUAL, symmetrical; regular; of the same number; like another in all respects, or in length.

Ē'QUALLȲ - PĬN'NĀTE, see ABRUPTLY-PINNATE.

ĒQUÀTŌ'RĬAL PLĀNE, the plane which passes through the equatorial plate (mother-star) of the cell-nucleus, or between the parts when the plate has divided, and which occupies the position of the future cell-wall. It is the plane of cell-division.

ĒQUÀTŌ'RĬAL PLĀTE, see MoTHER-STAR.

ĒQUĬLĂT'ĔRAL, equal-sided; opposed to Oblique.

ĒQUĬNŌC'TIAL, said of flowers which open and close at particular hours of the day.

ĔQ'UĬTANT, having the leaves so arranged that the base of each is enclosed within the opposite base of that which is next below it, as in the iris. Compare HALF-EQUITANT.

ĒQUĬVĂL'VŪLAR, having the valves of a capsule all of the same size.

ĒQUĬV'ŌCAL ĠĔNĔRĀ'TION, see SPONTANEOUS GENERATION.

ĒRĂDĬC'ŪLŌSE, without rootlets or rhizoids.

ĒRĔCT', perpendicular, or nearly so, to the surface to which it is attached; standing without support, not weak or lax. Compare STRICT. Applied to ovules or seeds it means growing vertically from the base of the ovary. Compare ASCENDING and INVERTED.

ĒRĔCTŌPĂT'ENT, intermediate between erect and spreading.

ĔRĒ'MŌBLĂST, a unicellular plant. (Sachs.)

ĔRĬĂN'THOŬS, woolly-flowered.

ĔRĬCĀ'ÇEOŬS, heath-like.

ĔR'ĬCOĪD, ericaceous. Said of a subulate form of leaves often found upon the juniper.

ĔRĬŌPHȲL'LOŬS, woolly-leaved.

ĒRŌ'DĔD, see EROSE.

ĒRŌSE', having irregular sinuses as if bitten out; eroded.

ĒRŌS'TRĀTE, without a beak.

ĒRŬM'PENT, breaking out, as the spore-clusters of some fungi through the epidermis of their host.

ĒRȲTH'RŌPHȲL, red coloring matter in plants.

ĔS'CŪLENT, used for food by man.

ĒSĔP'TĀTE, without septa.

ĔSŌTĔR'ĬC, originating within the organism. Compare EXOTERIC.

ĔSSĔN'TIAL CHĂR'ĂCTĔR, a feature which distinguishes a plant or group of plants from all others; diagnostic character.

ĔSSĔN'TIAL ÔR'GANS, stamens and pistils.

ĔSTĪ'VAL, pertaining to summer; æstival.

ĔS'TĬVĀTE, to pass the summer in a dormant condition. Compare HIBERNATE.

ĔSTĬVĀ'TION, (1) the arrangement of the floral organs in the bud—usually written Æstivation; (2) passing the summer in a dormant condition.

ĒTÆ'RĬŌ, a term sometimes applied to such fruits as the raspberry and blackberry.

Ē'TĬŌLĀTĔD, blanched by exclusion of light. Compare CHLOROSIS.

EŬÇȲC'LĬC, applied by Braun to flowers having the members in each whorl equal in number and alternating with those in an adjoining whorl.

EŪ'PHȲLL, an ordinary foliage-leaf. See PHYLLOME.

EŪTRŌP'ĬC, twining or turning with the sun; dextrorse.

ĒVĂLV'ŪLAR, without valves.

ĔVÅNĔS'ÇENT, (1) soon passing away; (2) applied to veins which disappear before reaching the margin of the leaf.

ĔVEN-PĬN'NĀTE, see ABRUPTLY-PINNATE.

ĔV'ĔRGREĒN, having green leaves throughout the year, as most Coniferæ.

ĔVĔRLÅST'ING FLOWĔRS, see IMMORTELLES.

ĒVĔR'NĪÆFÔRM, resembling the thallus of the lichen Evernia.

ĔVĔR'SION, the protrusion of a part which is generally produced in a cavity.

ĔVĔRT'ĔD, turned inside out.

ĔV'ĬDENT, distinctly visible.

ĒVĬT'TĀTE, without vittæ, which see.

ĔVŌLŪ'TION, the act of unfolding or unrolling; hence growth or development, especially the supposed development through successive generations of the higher from lower or simpler forms of life, both in animals and plants.

ĔXĂLBŪ'MĬNOŬS, having the nutriment in the seed all stored in the cotyledons.

ĔXĂN'NŪLĀTE, without an annulus.

ĔXĂR'ĬLLĀTE, without an aril.

ĔXĂS'PĔRĀTE, covered with short hard points; muricate.

ĔXÇĔN'TRĬC, out of the centre; abaxial. An embryo is *excentric* when it lies within the albumen, but not in the centre of it, as in asparagus; the trunk of a tree is *excentric* when developed more on one side of the heart than on the other.

ĔX'ÇĬPLE, see EXCIPULUM.

ĔX'ÇĬPŪLE, see EXCIPULUM.

ĔXÇĬP'ŪLŬM, the portion of the thallus supporting or surrounding the apothecium in lichens. In some cases the excipulum is an outer rim of the perithecium itself, and is then termed a "proper" excipulum.

ĔXÇĪTĂBĬL'ĬTY̆, the general faculty, characteristic of living bodies, of being influenced by external stimuli. Compare IRRITABILITY.

ĔXCRĔS'ÇENT, growing out in a morbid or unnatural manner, as a wart or tumor; superfluous.

ĔXCRĒ'TION, the separation of unassimilable matter from an organism. Compare SECRETION.

ĔXCŪR'RENT, projecting beyond the usual limit, or to the extreme summit or apex.

ĔXFŌ'LĬĀTE, to cast off layers or plates, as the bark of sycamore.

ĔXHÅLĀ'TION, see TRANSPIRATION.

ĔXĬG'ŪOŬS, small or slender. See GRACILE.

ĔX'ĬNE, see EXTINE.

ĔXĬN'TĬNE, a term applied by Fritsche to a third coat observed by him in the covering of certain pollen-grains between the intine and a second coat called by him the intexine. The terms Intexine and Exintine are not in ordinary use and do not represent any constant recognized structures.

ĔX'ŌCÄRP, the outer layer of a pericarp.

ĔXŎG'ĔNOŬS, growing by addition to the outside, or springing from the exterior tissues. Applied also to the manner of growth of the stem in ordinary trees (dicotyledons and gymnosperms).

ĔXŎG'Y̆NOŬS, having the style exserted beyond the corolla.

ĔXŌPĔRĬD'ĬŬM, the outer peridium when there are more than one, as in Geaster.

ĔXŌRHĪ'ZȦ (pl. **Ĕxōrhī'zæ**), an old term for exogen. Compare ENDORHIZA.

ĔXŌRHĪ'ZAL, the manner in which the radicle of dicotyledons is developed in germination. (Rare.)

ĔX'ŌSMŌSE, the passage of gases or liquids through a closed membrane from within outward, or from the denser to the rarer fluid in the process of osmose.

ĔX'ŌSPŌRE, see EPISPORE.

ĔXŌSPŌ'RĬŬM, see EPISPORE.

ĔX'ŌSTŌME, the orifice in the outer coat of an ovule or seed, which with the endostome forms the foramen.

ĔXŌSTŌ'SĬS, any indurated protuberance.

ĔXŌTĔR'ĬC, having its cause or origin outside the organism. Compare ESOTERIC.

ĔXŌTHĒ'ÇĬŬM, the outer coat or epidermis of an anther. Compare ENDOTHECIUM.

ĔXŌT'ĬC, introduced from a foreign country.

ĔX'PLȦNĀTE, spread or flattened out. Applied to a part usually rolled or folded. Compare COMPLANATE.

ĔXSĔRT'ĔD, protruding beyond the margin of a receptacle, as stamens beyond the corolla, or a panicle of a grass above the leaf-sheath.

ĔXSĬCCĀ'TȦ (pl. **Ĕxsĭccā'tæ**), a dried herbarium specimen; exsiccate.

ĔX'SĬCCĀTE, see EXSICCATA.

ĔX'SĬCCĀTĔD, dried; especially, collected and dried for preservation as a botanical specimen.

ĔXSTĬP'ŪLĀTE, without stipules.

ĔXSŬC'COŬS, destitute of juice.

ĔXTĒ'RĬOR, when applied to the parts of a flower, means the same as Anterior, i.e., the side away from the axis; lower; outer.

ĔX'TĬNE, the outer coat of a pollen-grain. (Exine of Schacht.) Compare INTINE.

ĔX'TRȦ-ĂX'ĬLLĀRY̆, situated out of the axil.

ĔX'TRȦ-ÇĔL'LŪLAR, outside of a cell.

ĔX'TRȦ-FŌLĬĀ'ÇEOŬS, not situated upon or near the leaves, as *extra-foliaceous* prickles.

ĔXTRȦVĂG'ĬNAL, applied to branches in grasses which in growth burst through the base of the subtending sheath. Compare INTRAVAGINAL.

ĔXTRŌRSE', applied to anthers the lobes of which are situated on the outside of the filament or connective, i.e., on the side farthest removed from the pistil. Such anthers generally dehisce on the outside also.

ĔXŬNGUĬC'ŪLĀTE, without an unguis or claw, as most petals.

ĔXŪ'VĬÆ, anything excreted or cast off. (Rare.)

EY̆E (Hort.), a bud on a tuber; the cavity enclosed by the calyx in the apple; the ostiolum or opening in the apex of a fig; any conspicuous central spot in a flower or petal, including the disk in Compositæ.

FĀÇE, the upper, inner, or free surface of an organ as opposed to the back.

FĀ'ÇĬĒŞ, the general aspect of a plant. (Obs.) Compare HABIT.

FĂC′ULTĀTĬVE, occasional; incidental. Compare OBLIGATE.

FĂC′ULTĀTĬVE PĂR′ĂSĪTE, a plant (usually a saporophytic fungus) capable of passing through at least certain stages of its development as a parasite, but which does not always or necessarily do so.

FĂC′ULTĀTĬVE SĂP′RŌPHĪTE, a plant (usually applied to parasitic fungi) which is capable of living as a saprophyte during the whole or a part of its life.

FÆC′ŪLÀ, see FECULA.

FĂL′CĀTE, scythe-shaped, or sickle-shaped. Compare UNCINATE.

FĂL′ÇĬFÔRM, see FALCATE.

FẠLSE, similar in appearance, but different in structure or origin; spurious. The same as the Greek pseudo-.

FẠLSE DĪCHŎT′ŌMY̆, any dichotomous appearance which does not arise from a terminal division of the main axis, as a dichasium.

FẠLSE DĬSSĔP′ĬMENT, one of the additional partitions in certain fruits which is not formed by the edges of carpels. False dissepiments frequently proceed from the dorsal suture.

FẠLSE ĬNDŪŞĬŬM, a recurved margin of the frond in ferns covering the sporangia, as in the genus Pteris.

FẠLSE RÀCĒME′, see HELICOID CYME.

FẠLSE PÀRĔN′CHY̆MÀ, see PSEUDOPARENCHYMA.

FĂM′ĬLY̆, the same as ORDER, which see, and in more familiar use. The term is also employed, especially among cryptogams, to indicate groups lower than the order. In horticulture it is sometimes used to indicate groups of related varieties, as the Duchess family among apples.

FĂN′-SHĀPED, like a fan in outline, especially if also plaited; flabelliform; flabellate.

FĂN′-VEINED, see PALMATELY-VEINED.

FÄRC′TĀTE, without vacuities; stuffed; obstructed; infarctate; infarcted; opposed especially to fistulose. Seldom used, the word solid or turgid being nearly always preferable. See STUFFED.

FÀRĪ′NÀ, starch. Formerly applied also to pollen.

FĂRĬNĀ′ÇEOŬS, containing starch, or of the texture of meal or flour.

FĂR′ĬNŌSE, covered with a white mealy powder.

FĂR′ĬNŌSE, n., a supposititious cellulose substance in starch-grains, which is not colored blue by iodine. Compare GRANULOSE.

FĂS′ÇĬÀ [*făsh-ĭ-ȧ*] (pl. **Făs′çĭæ**), a cross-band, especially of color. (Rare.)

FĂS′ÇĬĀTĔD, (1) having broad parallel bands or stripes; (2) exhibiting fasciation, which see.

FĂSÇĬĀ′TION, a monstrous flattened expansion of the stem, as in the garden cockscomb (Celosia).

FĂS′ÇĬCLE, a bundle, as the clustered leaves on the dormant branches or spurs of the larch; a bundle of tuberous roots, as in the dahlia; a fibrovascular bundle, especially if rudimentary; a close cyme, as in sweet-william; a bundle of herbarium specimens.

FĂS'ÇĬCLED, growing in tufts or clusters; fascicular; fasciculated.

FĂSÇĬC'ŪLAR, see FASCICLED.

FĂSÇĬC'ŪLAR SY̆S'TĔM, see FIBROVASCULAR SYSTEM.

FĂSÇĬC'ŪLAR TĬS'SŪE, see FIBROVASCULAR TISSUE.

FĂSÇĬC'ŪLĀTE, see FASCICLED.

FĂSÇĬC'ŪLĀTĔD, see FASCICLED.

FĂSÇĬC'ŪLŬS (pl. **Făscĭc'ūlī**), a dense cymose inflorescence, as in sweet-william; fascicle.

FĂSTĬĠ'ĬĀTE, having the branches close, parallel, and upright, as in Lombardy poplar. Sometimes erroneously used for flat-topped.

FAUX (pl. **Fau'çēs**), the throat or orifice of a gamopetalous or gamosepalous flower.

FȦVĔL'LȦ (pl. **Fȧvĕl'læ**), a form of sporocarp in Florideæ, consisting of an irregular mass of spores embedded in more or less gelatinous material and without a distinct conceptacle. It may be seated upon the frond or more or less embedded in it, and is derived from one or from several contiguous cells. In the latter case, and also when embedded in the frond, it was formerly called a favellidium.

FȦVĒ'ŌLĀTE, see ALVEOLATE.

FȦVŌSE', see ALVEOLATE.

FĔATH'ĔR-VEINED, see PINNATELY-VEINED.

FĔATH'ĔRY̆, see PLUMOSE.

FĔC'ŪLȦ, any powdery farinaceous matter.

FĔC'ŪLENT, muddy; thick with sediment.

FĒCŬNDĀ'TION, see FERTILIZATION.

FĒCŬN'DĬTY̆, fertility; fruitfulness.

FEED'ĔR, an outgrowth of the hypocotyl in the embryo of some genera of Gnetaceæ which serves for the absorption of the endosperm.

FĔLT'ĔD-TĬS'SŪE, hyphal tissue in which the filamentous cells are not regularly united, as in phenogams, but cross one another irregularly, and are often more or less grown together; tela contexta; spurious tissue. In its more consolidated forms it is known as pseudo-parenchyma.

FĒ'MĀLE FLOWĔR, one having pistils only; pistillate flower.

FĔNĔS'TRĀTE, having rather large openings like windows.

FĒ'RAL, see WILD.

FĔRRU'ĠĬNOŬS, resembling iron-rust; brownish-red. For synonyms see RUBIGINOSE.

FĔR'TĬLE, producing fruit, or reproductive bodies of any kind; having pistillate or perfect flowers.

FĔRTĬLĬZĀ'TION, the process by which the pollen causes the ovule to develop as a seed. It is the essential feature of sexual reproduction of every kind, being the union of the male and female reproductive bodies. In some cases, and perhaps always, it consists in the coalescence of the nuclei of two cells of different nature and origin; fecundation; impregnation. See CONJUGATION.

FEŬ'ILLEMÔRT, of the color of a faded leaf; filemot. (Rare.)

FĪ'BER, see FIBRE.

FĪ'BRĔ, any slender thread-like body of considerable strength; especially: (1) the slender fusiform cells of the inner bark, known as bast; (2) small slender roots like those of grasses. Also applied to bodies of similar form which have no special strength, as the threads or filaments in a nucleus during cell-division.

FĪ'BRĬL, diminutive of Fibre; a small or secondary fibre.

FĪBRĬL'LĂ (pl. Fībrĭl'læ), see FIBRIL.

FĪ'BRĬLLŌSE, diminutive of Fibrous; bearing fibrils or composed of small fibres.

FĪ'BRĬLLŌSE MȲCĒ'LĬŬM, see FIBROUS MYCELIUM.

FĪ'BROŬS, composed wholly, or in large part, of fibres; separable into fibres.

FĪ'BROŬS MȲCĒ'LĬŬM, mycelium in which the hyphæ form by their union elongated branching strands; fibrillose mycelium; mycelial strand.

FĪ'BRŌ-VĀ'SAL BŬN'DLE, see FIBROVASCULAR BUNDLE.

FĪ'BRŌ-VĀ'SAL STRĬNG, see FIBROVASCULAR BUNDLE.

FĪBRŌVĂS'CŪLAR BŬN'DLE, one of the characteristic elements in the stem of all flowering plants and the higher cryptogams. Isolated fibrovascular bundles form the "fibres" in the so-called pith of a cornstalk, and the veins in leaves. Each bundle usually consists of two parts, xylem and phloem (which see), the whole often surrounded by a special layer of cells called the bundle-sheath.

FĪBRŌVĂS'CŪLAR CÔRD, a term applied by Strasburger to a fibrovascular bundle in monocotyledons, but not generally adopted.

FĪBRŌVĂS'CŪLAR ÇȲL'ĬNDĔR, a name given to the peculiar fibrovascular system in the stem of Lycopodiaceæ. Sometimes used in exogens, especially in roots, where it is generally called "central cylinder."

FĪBRŌVĂS'CŪLAR SȲS'TĔM, the fibrovascular tissues of a plant taken together. In exogenous trees it includes the veins of the leaves, and all the material of the stem and branches, except the pith, medullary rays, and outer bark.

FĬD'DLE-SHĀPED, see PANDURIFORM.

FĬL'ȦMENT, the stalk of a stamen supporting the anther.

FĬLȦMĔN'TOŬS, slender and thread-like, or composed of filaments.

FĬLȦMĔN'TOŬS MȲCĒ'LĬŬM, one composed of free hyphæ, which are at most loosely interwoven with one another, but without forming bodies of definite shape and outline; floccose mycelium.

FĬLȦMĔN'TOŬS SPŌR'ŌPHŌRE, see SIMPLE SPOROPHORE.

FĬLĀ'RĬOŬS, see FILAMENTOUS.

FĬL'ĒMŎT, see FEUILLEMORT.

FĬL'ĬCOĪD, fern-like.

FĬL'ĬFŌRM, thread-shaped; slender, round, and of equal thickness throughout. Compare CAPILLARY.

FĬL'ĬFÔRM ĂPPȦRĀ'TŬS, a homogeneous, strongly refractive, cellulose cap often found at the apex of each synergida, especially in monocotyledons.

FĬLĬPĔN'DŬLOŬS, hanging by a thread.

FĪ'LŌSE, terminating in a thread-like process.

FĬM'BRĬȦ, a fringe.

FĬM'BRĬĀTE, fringed; bordered by lax, slender processes, generally larger than hairs.

FĬM'BRĬCĀTE, see FIMBRIATE.

FĬMBRĬL'LĀTE, diminutive of Fimbriate; having a very small or fine fringe; fimbrilliferous.

FĬMBRĬLLĬF'ĔROŬS, see FIMBRILLATE.

FĬN'GĔRED, see DIGITATE.

FĬS'SĬLE, capable of being split or divided.

FĬS'SION, the division of an organ which is usually entire; that mode of cell-division in which the cell separates into two nearly equal portions.

FĬSSĬP'ȦROŬS, reproducing by spontaneous division into two parts.

FĬS'TŪLAR, see FISTULOSE.

FĬS'TŪLĬFÔRM, tubular.

FĬS'TŪLŌSE, hollow and cylindrical, or nearly so, as the stems of many grasses; fistular; fistulous. Used especially when the hollow is of considerable size, as in reeds.

FĬS'TŪLOŬS, see FISTULOSE.

FLȦBĔL'LĀTE, see FAN-SHAPED.

FLȦBĔL'LĬFÔRM, see FAN-SHAPED.

FLĂC'ÇĬD, unable to support its own weight. Compare LAX.

FLȦĠĔL'LȦ, pl., see FLAGELLUM.

FLĂĠ'ELLȦRY̆, pertaining to or caused by flagella, as the *flagellary* movements of certain zoöspores.

FLĂĠ'ĔLLĀTE, (1) bearing flagella; (2) flagelliform.

FLȦĠĔL'LĬFÔRM, long and supple like a whip-lash; flagellate.

FLȦĠĔL'LŬM (pl. **Flăġĕl'lȧ**), any slender flexible process or organ, as (1) a solitary long swinging process of protoplasm on certain zoöspores (a large cilium); (2) a similar appendage to the cells of many bacteria; (3) a young flexible shoot, especially a long trailing branch of a vine (sarment).

FLĂT (Hort.), in describing fruits, means flattened endwise (depressed).

FLȦVĔS'ÇENT, yellowish.

FLĀ'VOŬS, see FLAVUS.

FLĀ'VŬS, pure pale yellow; lemon-yellow. Compare LUTEUS.

FLĔSH'Y̆, enlarged and somewhat soft, as a tuber. Compare SUCCULENT.

FLĔX'ŪŌSE, zigzag; wavy; winding; flexuous.

FLĔX'ŪOŬS, see FLEXUOSE.

FLŌAT'ĬNG, see NATANT.

FLŎC'ÇĪ, pl., see FLOCCUS.

FLŎCCŌSE', covered with matted woolly hairs, especially if they fall away in tufts. Said of the perithecia of Erysiphe when the appendages are of equal diameter throughout, more or less tortuous, and end abruptly, or in a straight point (thus distinguished from "hooked" and "dichotomous)."

FLŎCCŌSE' MY̆CĒ'LĬŬM, see FILAMENTOUS MYCELIUM.

FLŎC'CŪLENT, see FLOCCOSE.

FLŎC'CŬS (pl. **Flŏc'çī**), any woolly hair or thread, or a tuft of such filaments.

FLŌ'RȦ, the aggregate of the species of plants of a country or region, or a book which describes them.

FLŌ'RAL, pertaining to a flower.

FLŌ'RAL DĬ'ÅGRĂM, a drawing showing the relative position of the parts of a flower.

FLŌ'RAL ĔN'VĔLŌPES, in ordinary plants the calyx and corolla, sometimes including bracts when developed in a special manner so as to surround a flower: in grasses the flowering glume and palet.

FLŌ'RAL GLŪME, used by Dr. W. J. Beal instead of flowering glume or lower palet.

FLŌ'RAL LĔAF, see BRACT.

FLŌRĔS'ÇENÇE, the opening of flowers; blossoming; anthesis.

FLŌ'RĔT, an individual flower of a head or cluster, especially in Compositæ; floscule; flosculus; floweret.

FLŌRĬF'ĔROŬS, flower-bearing.

FLŌRĬP'ÅROŬS, floriferous; sometimes used when a proliferous branch or flower bears additional flowers instead of stems and leaves.

FLŎS'CŬLAR, see FLOSCULOUS.

FLŎS'CŪLE, see FLORET.

FLŎS'CŪLŌSE, see FLOSCULOUS.

FLŎS'CŪLOŬS, composed of or bearing florets: applied mainly to heads of flowers in Compositæ when composed of tubular florets only. Compare SEMIFLOSCULOUS.

FLOW'ĔR, the part of a plant immediately concerned in the production of seed. A complete flower in ordinary plants consists of pistils, stamens, corolla, and calyx, of which the two former are essential to the production of seed. The parts of a flower are modified leaves.

FLOW'ĔR-BŬD, an unopened flower or cluster of flowers.

FLOW'ĔRĔT, see FLORET.

FLOW'ĔR-HĔAD, see HEAD.

FLOW'ĔRĬNG GLŪME, the organ in grasses formerly called the lower palet. It may subtend one flower or more. Dr. W. J. Beal proposes the more appropriate term Floral Glume.

FLŪ'ĬTANT, floating in or upon water. Compare NATANT.

FLŪ'VĬAL, see FLUVIATIC.

FLŪVĬĂT'ĬC, belonging to flowing water; fluvial; fluviatile.

FLŪ'VĬĂTĬLE, see FLUVIATIC.

FŌLD'ĔD, (1) said of leaves in vernation when the two halves are simply brought together forward; (2) (Hort.) when a narrow projection of the flesh of an apple extends into the cavity. (Warder.) Compare LIPPED.

FŌLĬĀ'ÇEOŬS, leaf-like; having leaves intermixed with the flowers, as a *foliaceous* spike; consisting of thin laminæ or layers; foliose.

FŌLĬĀ'ÇEOŬS THĂL'LŬS, the thallus in lichens when flat and leaf-like and attached by one or few points; frondose thallus. Compare CRUSTACEOUS THALLUS.

FŌ'LĬAĠE LĔAVES, ordinary green leaves, in distinction from those which are transformed into petals, scales, etc.

FŌ'LĬAR-TRĀÇE, see LEAF-TRACE.

FŌLĬĀ'TION, the act of leafing out; frondescence. Sometimes used erroneously for prefoliation.

FŌLĬF'ĔROŬS, bearing or producing leaves; foliiferous; foliiparous.

FŌLĬĬF'ĔROŬS, see FOLIFEROUS.

FŌ'LĬĬFŌRM, leaf-shaped

FŌLĬĬP'ÅROŬS, producing leaves or leaves only.

FŌ'LĬŌLĀTE, pertaining to leaflets, as *trifoliolate*—having three leaflets.

FŌ'LĬŌLE, a little leaf or leaflet. (Rare.)

FŌLĪ'ŌLŬM (pl. **Fōlī'ŏlà**), see FOLIOLE.

FŌ'LĬŌSE, (1) abounding in leaves; foliaceous; leafy; (2) having the nature or appearance of a leaf.

FŌ'LĬOŬS, see FOLIOSE.

FŌ'LĬŬM (pl. **Fō'lĭà**), a leaf.

FŎL'LĬCLE, a simple pod opening by the ventral suture only, as in the milkweed (Asclepias).

FŎLLĬC'ŪLĀTE, having follicles.

FŌLLĬC'ŪLŬS, see FOLLICLE.

FOOT, a basal protrusion of the fern-plant which maintains its connection with the prothallus. Also a similar base to the seta in mosses.

FOOT-STALK, the stem of a leaf, flower, or other organ. See PETIOLE, PEDUNCLE, PEDICEL, STIPE.

FŌRĀ'MĔN (pl. **Fōrăm'ĭnà**), any small aperture, especially that in the integuments of the ovule, at which fertilization is effected. Compare MICROPYLE.

FŌRĂM'ĬNĀTĔD, having small holes or perforations. Compare LACUNOSE.

FŌRÀMĬN'ŪLŌSE, pierced with very small holes; diminutive of Foraminated.

FÔR'ÇĬPĀTE, like a pair of pincers.

FŌRKED, having two or more main branches arising from nearly the same point; furcate. Compare BIFURCATED.

FÔRM, (1) nearly the same as Variation, which see; (2) one of the conditions or states when several regularly appear among plants of the same parentage, as the short-styled *form* in heterostyled species.

FŌRM'ÅTĬVE, concerned with or serving for growth, as *formative material* (starch, albuminoids, etc.), *formative tissue* (meristem).

FŌRM-ĠĒ'NŬS, a so-called genus constituted by similar form-species, as Botrytis in fungi; pseudo-genus.

FÔRM-SPĒ'ÇIĒȘ, a particular phase in the development of a protean organism, as the rusts; so called because the different stages have often been mistaken for distinct species. Used also by E. L. Sturtevant synonymously with Race.

FŌRM-SPŌRE, a body which is morphologically or physiologically a spore, but which either does not become detached as an ordinary spore for dispersion, or which has not the power of germination.

FŌR'NĬCĀTE, see VAULTED.

FŌR'NĬX (pl. **Fôr'nĭceș**), arched scales in the throat of a corolla, as in comfrey.

FŎS'SĬL BŎT'ÅNY, the science of fossil plants, including their order of succession on the earth; paleobotany; paleophytology; geological botany; phytolithology.

FŌS'TĔR-PLĂNT, see HOST.

FOUR-FŌLD PŎL'LĔN-GRĀIN, see POLLEN-TETRAD.

FŌ'VĔÀ (pl. **Fō'vēæ**), a pit or depression, as that in the leaf of Isoetes, containing the sporangium.

FŌ'VĔĀTE, marked with deep or rather large pits or depressions. Compare ALVEOLATE.

FŌVĔ'ŌLÀ (pl. **Fōvĕ'ōlæ**), a little pit or depression; diminutive of Fovea.

FŌ'VĔŌLĀTE, marked with little pits or depressions.

FŌVĬL'LÀ (pl. **Fōvil'læ**), the contents of a pollen-grain; generally used in the singular for the entire contents, but sometimes applied to the individual granules in the fluid protoplasm.

FRĒĒ, not united with any dissimilar part; opposed to Adherent. Compare DISTINCT.

FRĒĒ ÇĔLL-FŌRMĀ'TION, the production of new cells within another, generally free from the cell-wall of the original or mother cell. Usually applied to cases in which several nuclei appear within the cell, each of which becomes surrounded by an independent cell wall, as in the production of pollen; internal or endogenous cell-formation.

FRĒĒ GRŌW'ĬNG, thrifty or vigorous, as opposed to dwarf or feeble.

FRĒĒ-STŎCK, a seedling tree of the same species used for grafting, as opposed to a stock of a different species, the latter being generally used for dwarfing.

FRĒĒ'STŌNE, applied to drupes in which the flesh separates readily from the pit when ripe. Compare CLINGSTONE.

FRĬLL, see ARMILLA.

FRĬNGED, see FIMBRIATE.

FRŎND, the leaf in ferns, especially the foliage portion of it; the expanded leaf-like portion of the thallus of liverworts; the stem and leaves taken together in plants where the distinction between leaf and stem is not obvious, or where, as in Equisetum, the leaves are unimportant; the whole expanded leaf-like or branching thallus of many marine algæ.

FRŎNDĔS'ÇENÇE, see FOLIATION and PHYLLODY.

FRŌN'DĬFŌRM, frond-like or frond-shaped, especially like the leaves of ordinary ferns.

FRŌN'DŌSE, frond-like, or producing fronds instead of ordidary foliage; leafy or leaf like. (Rare.) Compare THALLOID.

FRŌN'DŌSE THĂL'LŬS, see FOLIACEOUS THALLUS.

FRŎTH'Ÿ, see WARTY.

FRŬCTĔS'ÇENÇE, the time at which a fruit arrives at maturity.

FRŬCTĬFĬCĀ'TION, the fruit and attendant parts; an inflorescence at any stage of growth, the process of development of a fruit and its attendant parts.

FRŬCTĬFĬCĀ'TION, ŌR'GANȘ OF, stamens and pistils.

FRUIT, the mature ovary and its contents, together with any closely adhering part; specialized reproductive bodies of any kind, as the spores of cryptogams, including any modified portion of the plant in which they are produced. The term is also extended to many consolidated forms of inflorescence, as the cone of the pine.

FRUIT'-BŬD, generally the same as Flower-bud, which see.

FRUIT'-DŎT, see SORUS

FRUIT'-SPÛR, a short stout branch, bearing one or more flower-buds, as in the apple.

FRUMĔNTĀ'ÇEOŬS, producing or pertaining to edible grain.

FRŬS'TŪLE, the individual in Diatomaceæ (often joined together in colonies).

FRŬS′TŪLŌSE, consisting of similar separable parts, like the frustules of diatoms.

FRṲTĔS′ÇENT, somewhat shrubby,—woody at the base and herbaceous above, like the garden sage; subfrutescent; suffrutescent.

FRṲ′TĔX, a shrub, which see.

FRṲ′TĬCŌSE, shrubby; pertaining to shrubs; shrub-like. Compare FRUTESCENT.

FRṲ′TĬCŌSE THĂL′LŬS, a thallus in lichens which is attached to the substratum by one point only, or by a narrow base, and grows upward as a simple, or more usually branched, shrub-like body.

FRṲTĬC′ŪLŌSE, like a small shrub; diminutive of Fruticose.

FRṲTĬC′ŪLŬS, a little shrub.

FŪ′COĪD, pertaining to, or resembling, Fucus, a genus of marine algæ.

FŪGĀ′ÇIOŬS, disappearing in a very short time; ephemeral. Compare CADUCOUS.

FŪ′ĠĬTĬVE, quickly disappearing; easily blown away or absorbed; volatile; evanescent; fugacious.

FŬL′CRĀTE, furnished with fulcra, which see.

FŬL′CRŬM (pl. **Fŭl′crȧ**), a general term for various appendages to the plant which serve for support or defence, as tendrils, spines, prickles, hairs, etc. Now little used.

FŪLĬĠ′ĬNŌSE, see FULIGINOUS.

FŪLĬĠ′ĬNOŬS, dark brown; sooty or smoky.

FṲLL, applied to double flowers in which all the stamens and pistils are transformed into petals; completely double.

FŬL′VĬD, see FULVOUS.

FŬL′VOŬS, yellow, mixed with gray and brown; tawny.

FŬL′VŬS, see FULVOUS.

FŪ′MŌSE, smoke-colored; brownish gray.

FŪ′MOŬS, see FUMOSE.

FŬNDȦMĔN′TAL ŌR′GANS, root, stem, and leaf.

FŬNDȦMĔN′TAL SY̆S′TĔM, all that portion of the substance of the higher plants which is not included in the fibrovascular and epidermal systems. Compare CELLULAR SYSTEM.

FŬNDȦMĔN′TAL TĬS′SŪE, pith, cortex, and medullary rays; ground-tissue.

FŬṈĠĬCĪ′DAL, destructive to fungi; antimycotic.

FŬṈ′ĠĬCĪDE, anything destructive to the life of a fungus.

FŬṈ′ĠĬFÔRM, mushroom-shaped.

FŬṈĠĬL′LĬFÔRM, diminutive of Fungiform.

FŬṈ′GOĪD, fungus-like.

FŬṈGŎL′ŌĠY̆, see MYCOLOGY.

FŬṈ′GŌSE, spongy in texture, like many fungi.

FŬṈ′GOŬS, produced by a fungus; pertaining to a fungus or to fungi, as a *fungous* disease. The substantive form "Fungus" is also used as an adjective.

FŬṈ′GŬSED, injured by a fungus. (Rare.)

FŪ′NĬCLE, see FUNICULUS.

FŪNĬC′ŪLAR CŌRD, see FUNICULUS.

FŪNĬC′ŪLŬS, the stalk of an ovule or seed by which it is attached to the placenta; funicular cord; umbilical cord; podosperm. In Nidulariæ a hyphal cord attaching the peridiolum to the inner surface of the wall of the peridium.

FŪNĬL'ĬFÔRM, like a rope or cord, as the roots of many endogens.

FŬN'NĔL-FŌRM, tubular, and gradually enlarging upward to a spreading border; infundibuliform. Compare ROTATE and CYATHIFORM.

FÛR'CĀTE, see FORKED.

FÛR'CĀTĔD, see FORKED.

FÛRCĀ'TION, division into two (rarely said of more) main branches. See BIFURCATION.

FÛRFŪRĀ'ÇEOŬS, covered with soft scales easily displaced; scurfy.

FŬR'RŌWED, see SULCATE.

FŬSÇĔS'ÇENT, slightly fuscous.

FŬS'COŬS, grayish brown.

FŪ'SĬFÔRM, spindle-shaped; enlarged, terete, and tapering toward each end, as the roots of some radishes.

FŪ'SŌĬD, somewhat fusiform.

GĂL'BŪLŬS, a closed fleshy cone, resembling a berry, as that of juniper.

GĀ'LĔȦ, an arched sepal or petal resembling a helmet; helmet; hood; cucullus Also applied to the upper lip of some Scrophulariaceæ, though not so arched.

GĀ'LĔĀTE, helmet-shaped, as the flower of Aconitum.

GĂLVȦNŌT'RŌPĬṢM, the curvature of growing organs under the influence of a current of electricity.

GĂM'ĒTE, any sexual protoplasmic body, naked or invested with a membrane, motile or non-motile, as an oösphere or antherozoid; conjugation-cell; generative cell. Used mainly in Conjugatæ.

GĂMĒTŌĠĔN'ĔSĬS, the production of gametes (male or female).

GĂM'ĒTŌPHȲTE, the prothallus or sexual generation in ferns, etc. Compare SPOROPHYTE.

GĂMŌĠĔN'ĔSĬS, sexual reproduction.

GĂMŌPĔT'ALOŬS, having the petals more or less united; sympetalous. Also called erroneously Monopetalous. Compare POLYPETALOUS and UNIPETALOUS.

GĂMŌPHȲL'LOŬS, a term including Gamopetalous and Gamosepalous, but applied mainly to sepals.

GĂMŌSĔP'ALOŬS, having sepals more or less united; monosepalous.

GĂṈ'GLĬŎN (pl. Găn'glĭȧ), a term applied to various enlargements on the mycelium of certain fungi, some of which at least are rudimentary fructifications.

ĠEĪTŌNŎG'ȦMȲ, the fertilization of a pistil by pollen from another flower of the same plant—the closest kind of cross-fertilization.

ĠĔM, see GEMMA.

ĠĔM'ĬNĀTE, in pairs or twins; binate.

ĠĔM'MȦ (pl. Ġĕm'mæ), an old term for leaf-bud, now usually confined to various asexual reproductive bud-like processes in cryptogams. They may be distinguished from gonidia by not having as uniform methods of production, by greater variation in size, and by usually containing many cells. See GONIDIUM.

ĠĔMMĀ'ÇEOŬS, having the nature of gemmæ; bearing gemmæ; gemmiferous.

GĔMMĀ'TION, reproduction by means of gemmæ.

GĔMMĬF'ĔROŬS, bearing gemmæ.

GĔMMĬP'ÀROŬS, producing gemmæ.

GĔM'MŪLE, diminutive of Gemma—an old term for leaf-bud and plumule. Now applied to certain primary formative granules in the protoplasm. (Nägeli.)

GĔNĒAĠĔN'ĔSĬS, see PARTHENOGENESIS.

GĔN'ĔRÀ, pl., see GENUS.

GĔN'ĔRAL, see COMMON.

GĔN'ĔRAL ĬN'VŌLŪCRE, see COMMON INVOLUCRE.

GĔN'ĔRĀTĬNG TĬS'SŪE, see MERISTEM.

GĔN'ĔRĀTĬVE ÇĔLL, a sexual reproductive cell of any kind. See GAMETE. Also applied to the cell in a pollen-grain which develops into the pollen-tube. Compare VEGETATIVE CELL.

GĔN'ĔRĀTĬVE NŪ'CLĔŬS, the nucleus in the pollen-tube which is directly concerned in fertilization.

GĔNĔR'ĬC, pertaining to a genus.

GĔNĔT'ĬC, pertaining to generation or origin; e.g., things are *genetically* related which have the same origin.

GĔNĔT'ĬC SPĪ'RAL, a spiral line passing through the point of insertion of all equivalent lateral members on an axis from older to younger; generating spiral; fundamental spiral.

GĔNĬC'ŪLĀTE, bent abruptly at an angle, like the knee, as the stems of decumbent grasses.

GĔNĬC'ŪLŬM, a term occasionally applied to a node, especially when the stem is bent at that point, as is frequent in grasses.

GĔNŪFLĔC'TION, the formation of a knee-like bend in a conjugating filament, as in Sirogonium.

GĒ'NŬS (pl. Gĕn'ĕrà), a group of species within a family or order.

GĒ'NŬS-HȲ'BRĬD, a hybrid between plants of distinct genera; bigener.

GĒ'ŌBLĀST, a plumule which in germination leaves the cotyledons under ground, as in the pea.

GĒŌGRĂPH'ĬCAL BŎT'ÀNȲ, the study of plants in respect to their geographical distribution; botanical geography.

GĒŌLŎĠ'ĬCAL BŎT'ÀNȲ, see FOSSIL BOTANY.

GĒŎT'RŌPĬȘM, the tendency to grow downward or toward the centre of the earth. Compare APOGEOTROPISM.

GĔRM, a bud or growing point; the embryo in a seed; a rudimentary ovary or young fruit; a female reproductive cell—germ-cell, oösphere; a spore or seed; especially a spore or reproductive individual in bacteria.

GĔRM-ÇĔLL, any female reproductive cell. Compare SPERM-CELL. See OÖSPHERE. Applied also by Brefeld to spores of the simplest character (Sporidia) borne on a promycelium.

GĔR'MĔN, an old name for ovary.

GĔR'MĬNAL ĂPPÀRĀ'TŬS, see EGG-APPARATUS.

GĔR'MĬNAL CÔR'PŬSÇLE, see OÖSPHERE.

GĔR'MĬNAL VĔS'ĬCLE, see OÖSPHERE.

GĔRMĬNĀ'TION, the early stage of growth of a seed or spore into a new plant; sprouting.

GĔR'MĬNĀTĬVE NŬ'CLĔŬS, see GENERATIVE NUCLEUS.

GĔRM-NŬ'CLĔŬS, the nucleus resulting from the union of the pronuclei of two gametes in conjugation or fertilization.

GĔRM-PŌRE, a pit in the coat of a spore through which the germ-tube issues in germination.

GĔRM-TŬBE, the first growth from a spore or sclerotium upon germination.

GĬB'BOŬS, convex, as though swollen; protuberant, especially upon one side, or some distinct part of the surface.

GĬLLS, the spore-bearing plates upon the lower side of the cap in mushrooms; lamellæ.

GIR'DLE, the overlapping edge of one of the two valves in diatoms.

GLĀ'BRĀTE, nearly glabrous.

GLÅBRĔS'ÇENT, slightly glabrous.

GLÅBRĬŬS'CŪLŬS, almost but not quite glabrous.

GLĀ'BROŬS, smooth; free from roughness or hairs—the surface may be uneven. Compare SCABROUS and LÆVIS.

GLĂD'ĬĀTE, see ENSIFORM.

GLĂND, any secreting apparatus. A gland is generally a group of cells having a peculiar form and character to adapt them to their special function. They sometimes form wart-like projections upon the surface, or depressions within it. The hairs of many plants also serve as glands (see GLANDULAR HAIR). In deeply-seated glands of certain kinds, as those of the pine, the internal cell-walls of the gland are more or less absorbed to form reservoirs for the secreted substance. The term gland is also applied to certain wart-like swellings which are not secretory, as the abortive teeth at the base of the leaf in the peach and cherry.

GLĂN'DĬFÔRM, gland-shaped or gland-like; adenoid.

GLAND OF THE TORUS, see LEPAL.

GLĂN'DŪLAR, gland-like or bearing glands; glanduliferous.

GLĂN'DŪLAR DĬSK, see RETINACULUM.

GLĂN'DŪLAR HÂIR, an epidermal appendage of one or more cells, the apex of which is usually enlarged and contains the peculiar secretion.

GLĂN'DŪLAR WOŎD'Ў TĬS'SŪE, a term formerly applied to the woody tissue of Coniferæ from the appearance of its circular bordered pits.

GLĂN'DŪLĀ'TION, the position and arrangement of the glands upon a plant.

GLĂNDŪLĬF'ĔROŬS, bearing glands.

GLĂN'DŪLŌSE, see GLANDULAR.

GLĂN'DŪLŌSE-SĔR'RĀTE, having serratures tipped by so-called glands, as the leaves of *Prunus glandulosa*.

GLĂNS, a nut like that of the oak and chestnut, and sometimes extended to all large nuts. A term of little use.

GLÂR'ĔŌSE, growing in gravelly places.

GLAUÇĔS'ÇENT, slightly glaucous.

GLAU'COŬS, covered with a whitish bloom, as the leaves of cabbage; more accurately, light bluish green; sea-green. Compare PRUINOSE, HOARY, and CANESCENT.

GLĔ'BÀ (pl. **Glē'bæ**), chambered sporogenous tissue within a sporophore, as in puff-balls.

GLŌ'BĀTE, globular.

GLŌ'BOIDS, granules of calcium-magnesium phosphate found in grains of aleurone.

GLŌ'BŌSE, see GLOBULAR.

GLŎB'ŪLAR, spherical or nearly so; globose.

GLŎB'ŪLE, the antheridium or male organ of Characeæ.

GLŌCHĬD'ĬĀTE, barbed like an arrow or fish-hook.

GLŌ'CHĬS, a barbed hair or bristle.

GLŌM'ĔRĀTE, collected into a close round head.

GLŎM'ĔRŪLE, a capitate cyme.

GLŪ'MÀ, see GLUME.

GLŪMĀ'ÇEOŬS, bearing or resembling glumes.

GLŪME, one of the outer floral envelopes in grasses. The term as now used includes the bracts which subtend a spikelet (empty glumes) and the lower of the two bracts subtending the individual flower (flowering glume).

GLŪMĔL'LÀ, an obsolete term which has been applied both to the palet and lodicule in grasses.

GLŪMĔL'LŪLÀ, see LODICULE.

GNAUR, a knot. (Obs.)

GNŌMŎN'ĬCAL, bent at right angles. (Obs.) See GENICULATE.

GŎB'LĔT-SHĀPED, see CRATERIFORM.

GŌNĬD'ĬŌPHŌRE, a stalk bearing a gonidium.

GŌNĬD'ĬŬM (pl. **Gōnĭd'ĭà**), a general term for nearly all asexual reproductive bodies in cryptogams. Also applied to the algal host of lichens. Compare SPORE and CARPOSPORE.

GŌNĬM'ĬÀ (sing. **Gōnĭm'ĭŭm**), a term of little importance originated by Nylander and used by Tuckerman and others for pale bluish green gonidia in lichens.

GŌNĬM'ĬC LĀYĔR, the gonidial layer in certain lichens. "Gonidial layer" is preferable.

GŎN'ĬMOŬS, gonidial as applied to the algal host of lichens; gonimic. (Rare.)

GŌN'ŌPHŌRE, a stalk elevating the stamens and pistils only.

GŌNŎPH'ŌRŬM, see GONOPHORE.

GŎN'ŌPLĂSM, in Peronosporeæ, the portion of the protoplasm of the antheridium which passes through the fertilizing tube and coalesces with the oösphere. (De Bary.)

GŌRGE, see THROAT.

GŎS'SȲPĬNE, cottony.

GRĂÇ'ĬLE, slender.

GRÀFT-HȲ'BRĬD, a plant, or portion of a plant, which is supposed to have been essentially modified through the influence of a graft.

GRĀIN, the seed or fruit of Gramineæ; any small seed.

GRĀINED, having grain-like tubercles or processes, as those on the flowers of dock (Rumex).

GRĂMĬNĀ'ÇEOŬS, pertaining to grasses; gramineous.

GRÀMĬN'ĒAL, see GRAMINACEOUS.

GRÀMĬN'ĒOŬS, see GRAMINACEOUS.

GRĂMĬNŎL'ŌGȲ, see AGROSTOLOGY.

GRĂN'ŪLÀ-GŌNĬM'ĬÀ, an old term for the gonidia of lichens.

GRĂN'ŪLAR, composed of grains or granules; covered with small tubercles.

GRĂN'ŪLĀTE, see GRANULAR.

GRĂN'ŪLE, any small grain-like body.

GRĂNŪLĬF'ĔROŬS, see GRANULAR.

GRȦVĔ'ŌLENT, having a strong unpleasant odor.

GREEN-LĀYĔR, see MESOPHLŒUM.

GRĒGĀ'RĬOŬS, the same as Social; also applied to the fruiting spots or sori of a parasitic fungus when they appear in groups upon the host. Compare CESPITOSE.

GRŌSSĬFĬCĀ'TION, the swelling of the ovary after fertilization.

GROUND-TĬS'SŪE, see FUNDAMENTAL TISSUE.

GRŌW'ĬNG POINT, see PUNCTUM VEGETATIONIS.

GRŌWTH-FŌRM, a vegetable structure marked by some easily recognized feature of growth, characterizing stages in the lives of plants which are not necessarily closely related, as a filamentous fungus.

GRŌWTH'-RĬNG, see ANNUAL RING.

GRUMŌSE', see GRUMOUS.

GRU'MOŬS, consisting of clustered grains or tubercles; grumose.

GUÄRD-ÇĔLLS, special epidermal cells, usually two in number, enclosing the opening of a stoma, and which have the power of altering their shape so as to increase or diminish the size of the opening.

GUÄRD'ĬAN-ÇĔLLS, see GUARD-CELLS.

GŪ'LAR, pertaining to the throat.

GŬM, a name applied to various viscid (not oily) secretions of amorphous character which either dissolve in water or merely swell in it, as cerasin, the characteristic element of cherry gum.

GŬM-PĂS'SĂĠE, a glandular intercellular passage containing gum.

GŬS'SĔT, an intercellular space, either filled or hollow, at an angle where more than two cells meet.

GŬT'TĀTE, covered with small dots, as though sprinkled with some colored fluid.

GŬT'TĬFĔR, a plant which produces gum or resin.

GŬTTĬF'ĔROŬS, yielding gum or resin.

GŬT'TŪLĀTE, resembling small drops of oil or resin.

ĠЎMNĂX'ŌNЎ, a monstrous condition in which the placenta protrudes from the ovary.

ĠЎMNŌBLĂS'TŬS, having the ovary superior. (Obs.)

ĠЎMNŌCÄR'POŬS, naked-fruited; having the fruit destitute of hairs (rare), or free from the perianth or other covering; in fungi, having the hymenium exposed when the spores are maturing. Compare ANGIOCARPOUS.

ĠЎMNŌSPĔR'MOŬS, having the seeds naked (not enclosed in a pericarp), as in Coniferæ.

ĠЎMNŎS'TŌMOŬS, said of the mouth of the sporangium in mosses, when destitute of a peristome.

ĠЎM'NŌSPŌRE, a naked spore—one not produced in a receptaele.

ĠЎNÆÇĬ'ŬM, see GYNŒCIUM.

GYNAN'DER, a plant having the stamens inserted on the pistil. (Rare.)

GYNAND'RIAN, see GYNANDROUS.

GYNAND'ROPHORE, a stalk supporting the stamens and pistils above the insertion of the corolla; gonophore.

GYNANDROS'POROUS, bearing both male and female spores; applied in Œdogonieæ to certain female plants which produce androspores.

GYNAN'DROUS, having stamens and pistils united.

GYNAN'THEROUS, having stamens converted into pistils.

GYNEÇI'UM, see GYNŒCIUM.

GYN'OBASE, an elevated portion of the receptacle supporting the ovary, as in geranium. Compare GYNOPHORE.

GYNOBA'SIC, having a gynobase. Also applied to styles which are attached to the base instead of the summit of the ovary.

GYNODIŒ'ÇIOUS, having only pistillate flowers on one set of plants and perfect flowers upon another set. Compare GYNOMONŒCIOUS and ANDRODIŒCIOUS.

GYNŒ'ÇIUM (pl. Gynœ'çia), the pistils of a flower taken together.

GYNOMONŒ'ÇIOUS, having perfect and pistillate flowers on the same plant but no staminate flowers. Compare GYNODIŒCIOUS and ANDROMONŒCIOUS.

GYN'OPHORE, the stalk of a pistil elevating it above the receptacle; carpophore; basigynium; podogynium; thecaphore. (Obs.) Compare GYNOBASE.

GYNOSTE'GIUM (pl. Gynoste'gia), a sheath or covering of the gynœcium, as the monadelphous filaments of Asclepias.

GYNOSTE'MIUM, a term formerly used for the united stamens and style (column) in orchids.

GY'RATE, coiled, circinate, or taking a circular course. Compare GYROSE.

GYRO'MA, an old term for the annulus of ferns.

GYROSE', curved alternately backward and forward; nearly the same as Anfractuose. Sometimes used in the sense of Gyrate.

HAB'IT, the general appearance or manner of growth, as loose or compact, dwarf or otherwise, climbing, creeping, or upright. Also the character with regard to fruitfulness, hardiness, etc.

HAB'ITAT, the kind of situation in which a plant is naturally found, as marsh, woods, mountains, etc. Compare HABITATION.

HABITA'TION, the entire locality or geographical range within which a species is found. Compare HABITAT and STATION.

HAD'ROME, see XYLEM. Applied by Ptonie to the phloëm-like portion of the fibrovascular bundle in vascular cryptogams.

HAIR, any feeble outgrowth from the epidermis; trichome. Hairs may be of any shape, and may consist of one cell or more. They are usually derived from a single epidermal cell.

HAIR-POINTED, terminating in a very fine weak point.

HAIR'Y, covered with longer and

coarser hairs than "pubescent."

HAL'BĔRD-SHĀPED, see HASTATE.

HAL'BĔRT-SHĀPED, see HASTATE.

HĂLF-ĂNĂT'RŌPOŬS, see AMPHITROPOUS.

HĂLF-BREED, applied in stock-breeding to a cross between a well-established breed and common or "scrub" stock, but seldom used in botany. (Used by Burbidge in the sense of Cross.)

HĂLF-ĔQ'UĬTANT, said of opposite leaves whose margins are folded forward and enclose the stem and one edge of the opposite leaf, leaving one margin of each leaf outside. Compare EQUITANT.

HĂLF-ĬNFĒ'RĬOR, said of an ovary when the stamens are perigynous.

HĂLF STĔM-CLĂSP'ĬNG, see SEMIAMPLEXICAUL.

HĂLF-SŪPĒ'RĬOR, see PERIGYNOUS.

HĂLŎPH'ĬLOŬS, salt-loving.

HĂL'ŌPHȲTE, a plant containing a large quantity of common salt in its composition, and which thrives best in salty places, as *Salsola Kali*.

HĂLVED, see DIMIDIATE.

HĀ'MĀTE, hooked.

HĀMŌSE', see HAMATE.

HĀ'MOŬS, see HAMATE.

HĂM'ŪLĀTE, diminutive of Hamate.

HĂM'ŪLŌSE, diminutive of Hamose; bearing small hooks.

HĂM'ŪLŬS, a small hook.

HĂPLŌGŌNĬD'ĬŬM, an algal gonidium in lichens resembling Protococcus. (Rare.)

HĂPLŌPĔRĬS'TŌMOŬS, having a peristome in mosses with but a single row of teeth.

HĂPLŌSTĔM'ŌNOŬS, having the stamens in one whorl.

HĂRD, said of fruits, chiefly pears, which require cooking to soften them for eating.

HĂRD'Ȳ, said of plants capable of passing the winter uninjured by cold. Hardiness also implies the ability to withstand any injurious climatic influence, but its limited use referring to cold is most common.

HĂS'TĀTE, like the head of a halberd—applied to leaves which have a spreading lobe on each side of the base. Compare SAGITTATE.

HĂS'TĬFÔRM, see HASTATE.

HĂS'TĬLE, see HASTATE.

HĂTCH'ĔT-SHĀPED, see DOLABRIFORM.

HAULM, the dead stems of any herbaceous plant.

HAUSTŌ'RĬŬM (pl. **Haustō'rĭă**), the special organ of certain parasites by means of which they obtain food from their host.

HĔAD, any compact somewhat rounded body upon a stem. The term is also applied to a cluster of nearly sessile flowers, as in the clovers and Compositæ, also to other more or less compact inflorescences, as the spike, corymb, and panicle. See CAPITULUM.

HEĂRT, the organic centre of anything, as the central portion of a tree-trunk, or a growing point surrounded by leaves.

HEĂRT-SHĀPED, see CORDATE.

HEĂRT-WOOD, see DURAMEN.

HĒBĒCĂR'POŬS, having pubescent fruit. (Obs.)

HĔB'ĒTĀTE, having an obtuse point; blunted.

HĔDĒRĀ'ÇEOŬS, pertaining to or resembling ivy.

HĔD'ĔRAL, see HEDERACEOUS.

HĔLĬÇ'ĬFÔRM, see HELICOID.

HĔL'ĬCŌID, (1) coiled into the form of a helix or snail-shell; spiral; (2) in inflorescence contrasted with Scorpioid, which see See HELICOID CYME and HELICOID DICHOTOMY.

HĔL'ĬCŌID ÇȲME, one in which each successive flower is situated upon the same side of a pseudaxis, which may or may not be coiled, as the primary branches of the inflorescence of *Hemerocallis fulva;* bostrychoid cyme; bostryx; false raceme. Compare SCORPIOID CYME.

HĔL'ĬCŌID DĬCHŎT'ŌMY̆, a dichotomy in which a branch on the same side in each successive bifurcation continues to develop while the other does not; bostrychoid dichotomy. Compare SCORPIOID DICHOTOMY.

HĒLĬŌG'Y̆'RĀTE, having a circular line carried obliquely around an object, as the annulus on the spore-case of Trichomanes.

HĒLĬŎT'RŌPĬSM, having the power of movement under the influence of light.

HĔL'MĔT, see GALEA.

HĔL'MĔT-SHĀPED, see GALEATE.

HĒLŌ'BĬOŬS, see PALUSTRINE.

HĔMĬ-ĂNĂT'RŌPOŬS, see AMPHITROPOUS.

HĔM'ĬCĂRP, one of the ripened separable carpels of a dicarpellary fruit, as in Umbelliferæ; diachænium. See MERICARP.

HĔM'ĬÇȲCLE, half of a coil or circle.

HĔMĬÇȲC'LĬC, having part of the floral organs arranged in whorls and the remainder in a spiral. Compare CYCLIC and ACYCLIC.

HĔMĬT'RŌPAL, see AMPHITROPOUS.

HĔMĬT'RŌPOŬS, see AMPHITROPOUS.

HĔPTĂG'Y̆NOŬS, having seven pistils or styles.

HĔPTĂM'ĔROŬS, having seven parts.

HĔPTĂN'DROŬS, having seven stamens.

HĔPTĂPĔT'ALOŬS, having seven petals.

HĔRB, a plant of which the stem contains but little wood and dies to the ground at the close of the season. It may be an annual, a biennial, or a perennial.

HĔRBĀ'ÇEOŬS, like an herb; succulent. Also green, as opposed to colored like an ordinary corolla, as a petal with an *herbaceous* (green) tip.

HĔRBĀ'ÇEOŬS PĔRĔN'NĬAL, see PERENNIAL HERB.

HĔRB'AL, see HERBARIUM.

HĔRBĀ'RĬŬM (pl. **Hĕrbā'rĭŭms** or **Herbā'rĭȧ**), a classified collection of dried specimens of plants; herbal; hortus-siccus.

HĔRBĔS'ÇENT, herbaceous or somewhat so.

HĔRBŌRĬZĀ'TION, see BOTANIZING.

HĔRCŎG'ĂMOŬS, said of an hermaphrodite flower when some structural obstacle prevents self-fertilization, as in many orchids.

HĔRMĂPH'RŌDĪTE, see PERFECT.

HĔSPĒRĬD'ĬŬM, a fruit like the orange, being succulent within and covered with an indehiscent leathery rind.

HĒTÆ'RĬŌ, a collection of distinct indehiscent carpels belonging to a single flower. They may be either dry upon a fleshy receptacle, as in the strawberry, or dry upon a dry receptacle, as in Ranunculus, or fleshy upon a dry receptacle, as in the raspberry. Usually but improperly spelled Eterio.

HĔTĔRAUXĒ'SĬS, any irregular or unsymmetrical growth, either normal or abnormal.

HĔTĔRŌCÄR'POŬS, bearing fruit of two or more kinds or forms, as in the genus Amphicarpea. Compare HOMOCARPOUS.

HĔTĔRŌÇĔPH'ÅLOŬS, bearing heads of more than one kind. For example, having flower-heads some of which contain only staminate flowers and some only pistillate.

HĔTĔRŌCHRŌ'MOŬS, having different members unlike in color; also applied to a flower-head in Compositæ when the florets of the centre or disk differ in color from those of the circumference or ray.

HĔTĔRŌCLĪ'NOŬS, having male and female flowers in separate heads or receptacles.

HĔT'ĔRŌÇЎST, one of the intercalated cells of special character in the filaments of Nostochineæ; limiting-cell. They are usually large, rounded, brownish, and glassy in appearance.

HĔTĔRŎD'RŌMOŬS, turning or coiling in opposite directions, as a tendril which coils first one way and then the other, or a plant on which the leaf-spiral of a branch runs in the opposite direction from that of the main axis. Compare HOMODROMOUS and ANTIDROMOUS.

HĔTĔRŒ'ÇIOŬS, parasitic on different plants at different stages of growth; metœcious; metoxenous; heteroxenous.

HĔTĔRŒÇĬŞ'MAL, see HETERŒCIOUS.

HĔT'ĔRŒÇЎST, see HETEROCYST.

HĔTĔRŎG'ÅMOŬS, said of the heads of flowers in Compositæ when the florets are not all alike in sex.

HĔTĔRŌĠĒ'NĔOŬS, not of uniform substance or character.

HĔTĔRŌĠĔN'ĔSĬS, see SPONTANEOUS GENERATION.

HĔTĔRŎG'ŌNOŬS, having two or more kinds of flowers differing in the relative lengths of the stamens and styles; heterostyled. See DIMORPHOUS and TRIMORPHOUS.

HĔTĔRŎM'ÅLOŬS, spreading in all directions. Compare HOMOMALOUS.

HĔTĔRŎM'ĔROŬS, having a different number of parts in the different whorls of a flower. Compare ISOMEROUS. Applied also to a lichen thallus when a layer of the algal cells divides it into an outer cortical and an inner medullary portion. Compare HOMOIOMEROUS.

HĔTĔRŌMÔR'PHOŬS, of two or more forms, as the flowers of *Buchloë dactyloides.*

HĔTĔRŌPHЎL'LOŬS, having two or more distinct sorts of foliage-leaves on the same plant, as in junipers; also applied to species whose leaves differ widely from those of related species. The term is not usually applied to plants in which the leaves merely assume different forms at successive elevations on the stem.

HĔTĔRŌRHĪ'ZAL, having roots which seem to proceed from no fixed point, as those of acrogens; said also of spores which germinate indifferently from any portion of the surface. Little used.

HĔTĔRŎS'PŌROŬS, bearing asexually produced spores of more than one kind, as in the Uredineæ; having macrospores and microspores, as in Selaginella. Compare HOMOSPOROUS and ISOSPOROUS

HĔT'ĔRŌSTȲLED, see HETEROGONOUS.

HĔT'ĔRŌTĂXȲ, the deviation of organs from their normal position.

HĔTĔRŎT'RŌPAL, see AMPHITROPOUS.

HĔTĔRŎT'RŌPOŬS, see AMPHITROPOUS. Also applied to any part which is turned in an unusual direction.

HĔTĔRŎX'ĒNOŬS, see HETERŒCIOUS.

HĔX-, a prefix derived from the Greek, meaning six. See SEX-.

HĔXĂG'ȲNOŬS, having six pistils or styles.

HĔXĂM'ĔROŬS, having the parts in sixes. Applied mainly to the parts of a flower, and meaning six organs in each whorl. Also written 6-merous.

HĔXĂN'DROŬS, having six stamens; hexastemonous.

HĔXȦPĔT'ALOŬS, having six petals.

HĔXȦPHȲL'LOŬS, having six leaves or leaflets.

HĔXȦSTĔM'ŌNOŬS, see HEXANDROUS.

HĪBĔR'NȦCLE, see HIBERNACULUM.

HĪBĔRNĂC'ŪLŬM, a protection for a growing part through the winter, as a bud or bulb.

HĪBĔR'NAL, pertaining to winter; blooming or vegetating in winter; hiemal; hyemal.

HĪBĔRNĀ'TION, passing the winter in a dormant condition.

HIDDEN-VEINED, having the veins of a leaf buried in the tissue so as not to be easily visible.

HIDE-BOUND, see BARK-BOUND.

HĪ'ĒMAL, see HIBERNAL.

HĪ'LAR, pertaining to the hilum.

HĪLE, see HILUM.

HĪ'LŬM (pl. **Hī'lȧ**, or preferably **Hī'lūms**), the scar, or point of attachment of a seed. The term is also applied to the nucleus of a starch-grain.

HIP, the fruit of the rose; a cynarrhodium.

HĬPPŌCRĔP'ĬFŌRM, horseshoe-shaped.

HĪRSŪTE', clothed with rather numerous long coarse hairs, harsher than pubescent and less harsh than hispid.

HĪR'TŬS, indefinite in meaning, but nearly the same as Hirsute, which see.

HĪRTĔL'LOŬS, slightly hirsute; stiffly pubescent.

HĬS'PĬD, clothed with erect stiff hairs, as Borage.

HĬSPĬD'ŪLOŬS, minutely hispid.

HĬSTĬŎL'ŌĠȲ, see HISTOLOGY.

HĬSTŌDĬĂL'ȲSĬS, the separation of the cells of a tissue from each other.

HĬSTŌĠĒNĔT'ĬC, tissue-forming; pertaining to histogeny.

HĬSTŌĠĔN'ĬC, see HISTOGENETIC.

HĬSTŎĠ'ĔNȲ, the origin or formation of tissue.

HĬSTŎL'ŌĠȲ, the science of the structure of tissues. Compare MORPHOLOGY.

HŌARY, grayish white; canescent.

HŌLD'-FAST, any root or root-like organ whose chief function is to retain the plant in place, as the aërial roots of ivy, or the suckers or rhizoids of many sea-weeds; crampon.

HŎLĒRĀ'ÇEOŬS, see OLERACEOUS.

HŎLŌCÄR'POŬS, having the pericarp entire. (Rare.)

HŎLŌSĂP'RŌPHȲTE, a complete saprophyte: one which lives entirely on dead organic matter.

HŎLŌSĒRĬÇ'EOŬS, covered with very short silky hairs hardly visible to the eye. Compare VELUTINOUS.

HŌMŌBLĂS'TĬC, said of the embryo when in its usual position with the radicle directed toward the micropyle and the cotyledons in the opposite direction. Compare ENANTIOBLASTIC.

HŌMŌCÄR'POŬS, bearing fruit all of one kind. Compare HETEROCARPOUS.

HŌMŌÇĔN'TRĬC, see CONCENTRIC.

HŌMŌCHRŌ'MOŬS, of uniform color.

HŌMŎD'RŌMAL, see HOMODROMOUS.

HŌMŎD'RŌMOŬS, turning continuously in the same direction. Compare HETERODROMOUS.

HŌMŌDȲ'NȦMOŬS, of equal strength, size, or vigor.

HŌMŎG'ȦMOŬS, having all the florets of a head in Compositæ alike in sex. Compare HETEROGAMOUS. Also used for Synacmic, which see.

HŌMŌĠĒ'NĒOŬS, having the same nature or structure throughout.

HŌMŎG'ŌNOŬS, having the stamens and pistils alike in character in all flowers of the species; homostyled. Compare HETEROGONOUS.

HŌMŌIŎM'ĒROŬS, applied to a lichen thallus in which the gonidia and hyphæ are mingled together and not distinctly stratified. Compare HETEROMEROUS.

HŌMŎL'ŌGOŬS, of the same morphological nature, as leaves, bracts, sepals, petals, stamens, and pistils are all *homologous*, or forms of the same fundamental organ.

HŎM'ŌLŎGUE, a part homologous with another, as a stamen with a leaf; homotype. Compare ANALOGUE. A nectary, for example, is in some cases the homologue of a stamen and in others of a petal, but it is always the analogue of any other nectary.

HŌMŎL'ŌĠȲ, correspondence in structure or morphological nature. Compare ANALOGY. See HOMOLOGOUS and HOMOLOGUE.

HŌMŎM'ȦLOŬS, applied to leaves or other organs which originate on the different sides of a stem, but are all turned toward one side. Compare SECUND.

HŌMŌMÔR'PHOŬS, of the same shape or character, as when the disk-flowers as well as the ray-flowers of a head in Compositæ are ligulate.

HŌMŌPĔT'ALOŬS (obs.), see REGULAR.

HŌMŎS'PŌROŬS, having asexually produced spores of only

one kind; isosporous. Compare HETEROSPOROUS.

HŌ'MŌSTȲLED, see HOMOGONOUS.

HŌMŌT'RŌPAL, see HOMOTROPOUS.

HŌMŎT'RŌPOŬS, said of an embryo in a curved seed when it is curved in the same manner as the seed.

HŎM'ŌTȲPE, see HOMOLOGUE.

HŌN'EY, see NECTAR.

HȮN'EY-CŌMBED, see ALVEOLATE.

HȮN'EY-DEW, a sweet substance found on the leaves of plants, usually a secretion from plant-lice.

HON'EY-GUĬDE, see NECTAR-GUIDE.

HON'EY-PŌRE, the supposed pore in flowers which secretes honey. (Obs.)

HȮN'EY-SPŌT, see NECTAR-GUIDE.

HO͝OD, see CUCULLUS.

HO͝OD'ED, see CUCULLATE.

HO͝OD'-SHĀPED, see CUCULLATE.

HO͞OP, the connecting band between the valves in Diatomaceæ.

HŌRĬZŌN'TAL SȲSTĔM, the cellular as distinguished from the fibrovascular system. Little used.

HŌR'MŌGŌN, see HORMOGONIUM.

HÔRMŌGŌNĪ'ŬM (pl. **Hôrmōgōnī'ȧ**), a reproductive body in certain algæ, as the Oscillatorieæ, consisting of a short chain of cells, one of the natural fragments of a filament.

HŌRN, any horn-shaped appendage, as the spur of a flower.

HŌRN'LĔT, a little horn.

HŌRNY, of the texture of a horn, as the pericarp of witch-hazel, *Hamamelis Virginica.*

HŌRŌLŎG'ĬCAL, said of flowers which open and close at definite hours of the day.

HŌRTĔN'SĬS, pertaining to a garden.

HÔR'TŬS-SĬC'CŬS, see HERBARIUM.

HŌ̤SE-IN-HŌ̤SE, when the calyx in a gamopetalous flower takes the form of the corolla, or when the corolla itself in such a flower is in two parts or whorls.

HŎST, a plant which supports a parasite.

HŌST'-PLĂNT, see HOST.

HŪ'MĬFŪSE, spreading upon the ground.

HŪ'MĬLĬS, low, or less in stature than related species, but not necessarily dwarf; pumilus.

HŪ'MŬS PLANT, see SAPROPHYTE.

HŬSK, any large, dry, thin envelope covering the fruit or inflorescence, as one of the bracts surrounding an ear of corn.

HȲȦLĔS'ÇENT, somewhat hyaline.

HȲ'ȦLĬNE, clear and colorless like glass or water; translucent or transparent.

HȲ'ȦLŌPLĂ̤SM, the clear portion of the protoplasm free from granules. Often restricted to such a layer next to the cell-wall, then called by some Ectoplasm.

HȲBĔR'NȦCLE, see HIBERNACULUM.

HȲ'BĔRNĀTĬNG, see HIBERNATING.

HȲ'BRĬD, the offspring of two species of the same genus. Compare CROSS. The term

Hybrid is often erroneously used to designate the result of cross-fertilization between any different species or varieties. As true hybrids are often sterile, the term "mule" has been applied to them. Burbidge proposes to retain the term "mule" for sterile hybrids only. See GENUS-HYBRID.

HȲBRĬDĪZĀ'TION, the fertilization of a flower by pollen from a plant of another species. Often erroneously used for cross-fertilization in general.

HȲ'DROID, see TRACHEID.

HȲDRŎPH'ĬLOŬS, having the pollen conveyed to the stigma by means of water.

HȲ'DRŌPHȲTE, an aquatic plant of any kind.

HȲDRŎT'RŌPĬSM, power in a growing organ of turning in a definite manner or direction through the influence of moisture, i.e. of taking a definite position with respect to the source of moisture. See POSITIVE and NEGATIVE HYDROTROPISM. Compare HYGROSCOPIC.

HȲĒ'MAL, see HIBERNAL.

HȲGRŌMĔT'RĬC, moving in a definite manner as a result of a change in the degree of moisture; hygroscopic.

HȲGRŎPH'ÀNOŬS, having a watery appearance.

HȲ'GRŌPLĂSM, the fluid portion of the protoplasm. Compare STEREOPLASM.

HȲGRŌSCŎP'ĬC, (1) absorbing moisture with avidity; (2) showing an increase or diminution of moisture by motion; hygrometric. Compare HYDROTROPISM.

HȲGRŌSCŎP'ĬC ÇĔLLS, certain cells in the leaves of grasses which have the power of altering their form under the influence of moisture and causing the leaves to "curl" in dry weather. From their bladder-like appearance they are also called Bulliform Cells.

HȲMĒ'NĬŬM (pl. **Hȳmē'nĭà**), a spore-bearing surface in fungi, especially in mushrooms and their allies.

HȲMĔN'ŌPHŌRE, the portion of a sporophore immediately beneath a hymenium; hymenophorum.

HȲMĒNŎPH'ŌRŬM, see HYMENOPHORE.

HȲPĂN'THĬŬM, an expanded, usually fleshy, receptacle, more or less enclosing the flowers, as in the fig, rose, Dorstenia and Ambora; hypanthodium. See HIP and SYCONUS. Compare CLINANTHIUM and RECEPTACULAR TUBE.

HȲPĂNTHŌ'DĬŬM, see HYPANTHIUM.

HȲPĔRBŌ'RĒAN, growing in the extreme north.

HȲPĒR'TRŌPHȲ, excessive development. Compare ATROPHY.

HȲ'PHÀ (pl. Hy'phæ), a filament of mycelium.

HȲPHĂS'MÀ, an old term for mycelium, still occasionally used when particularly delicate and web-like.

HȲP'NŌSPĔRM, an asexually produced resting-spore in algæ; hypnospore.

HȲPNŌSPŌRĂN'ĠĬŬM, a sporangium containing resting-spores.

HȲP'NŌSPŌRE, any resting-spore, especially one produced asexually. Compare HYPNOSPERM. See RESTING-SPORE.

HYPO-, in Greek derivatives, under.

HYPOCARPOGE'AN, producing fruit beneath the surface of the ground, as the peanut.

HYP'OCHIL, see HYPOCHILIUM.

HYPOCHIL'IUM, the lower or basal part of the divided labellum in certain orchids; hypochil. Compare EPICHILIUM.

HY'POCOTYL, the caulicle. Compare EPICOTYL.

HYPOCOTYLE'DONARY, situated below the cotyledons.

HYPOCRATER'IFORM, having a long tube with an abruptly spreading border, as in phlox; salver-form.

HYP'ODERM, see HYPODERMA.

HYPODER'MA, cells or layers of cells next beneath the epidermal system which are developed in a special manner, usually as collenchyma or other strengthening tissue.

HYPODER'MAL, situated beneath the epidermis; hypodermous.

HYPODER'MOUS, see HYPODERMAL.

HYPOGÆ'AN, see HYPOGEAN.

HYPOGÆ'OUS, see HYPOGEAN.

HYPOGE'AL, see HYPOGEAN.

HYPOGE'AN, subterranean: applied to parts which grow beneath the surface of the ground, and to plants which ripen their fruit beneath the surface; hypogæan; hypogæous; hypogeal; hypogeous.

HYPOG'ENOUS, growing upon the lower surface of anything.

HYPOGE'OUS, see HYPOGEAN.

HYPOG'YNOUS, growing beneath the pistil, and free.

HYPONAS'TIC, denoting curvature from growth on the lower side of an organ, causing it to bend upward. Compare EPINASTIC.

HY'PONASTY, having more rapid growth upon the lower than upon the upper surface. Compare EPINASTY.

HYPOPHLŒ'ODAL, beneath the bark.

HYPOPH'YLLOUS, situated upon the lower side of a leaf.

HYPOPHYL'LUM, an abortive or scale-like leaf subtending anything. (Rare.)

HYPOPH'YSIS, an appropriate but seldom used term for the Apophysis in mosses.

HYPOTHAL'LUS, a lower or interior stratum in a thallus.

HYPOTHE'ÇIUM, a portion of the thallus beneath or around the apothecium in lichens.

HYP'SOPHYLL, see BRACT.

HYSTERAN'THOUS, said of plants which have the flowers expand after the leaves have appeared. The leaves therefore in a hysteranthous plant are proteranthous.

HYSTEROGEN'IC, formed late; applied to intercellular spaces formed in older tissues. Compare PROTOGENIC.

ICOS-, in Greek compounds, twenty.

ICOSAN'DROUS, having twenty or more perigynous stamens. Compare POLYANDROUS.

IDENTIFICA'TION, see DETERMINATION.

ID'IOBLAST, a single cell in a tissue which differs greatly from its neighbors in form, size, nature of cell-wall, or cell-contents.

ID'IOPLASM, a term applied by Nägeli to the active organizing part of the protoplasm.

ĬG'NĒŬS, fiery-red, a lively scarlet.

ĬMBĔR'BĬS, without a beard or other hairs. See GLABROUS.

ĬMBĬBĬ'TION, the addition of moisture to organized bodies in a manner which causes them to swell up. Compare ABSORPTION.

ĬM'BRĬCĀTE, overlapping like tiles or shingles on a roof, as the scales of buds. In æstivation at least one piece is wholly external and one wholly internal.

ĬM'BRĬCĀTĔD, see IMBRICATE.

ĬM'BRĬCĀTĬVE, see IMBRICATE.

ĬMMÄR'ĠĬNĀTE, without a distinct rim or border.

ĬMMĔRSED', (1) growing wholly under water; demersed; submersed. (2) When one part or organ is completely embedded in another; innate. Compare EMERSED.

ĬMMŌRTĔLLEȘ', a term applied to various plants, the flower-heads of which retain their original shape and an attractive appearance when dried, as Gnaphalium and some other Compositæ; everlasting flowers.

ĬMPĂRĬPĬN'NĀTE, pinnate with a terminal leaflet, thus usually making an odd number; oddly pinnate.

ĬMPĔR'FĔCT, said of a flower which lacks either stamens or pistils.

ĬMPĔR'FŌRĀTE, closed; without an opening.

ĬMPRĔGNĀ'TION, see FERTILIZATION.

IN-AND-IN, breeding for successive generations from closely related individuals. Growing a "stock" or "strain" of corn on the same farm for many years would be called in-and-in breeding.

ĬNĀNE', empty.

ĬNĂN'THĔRĀTE, bearing no anther; said of certain sterile filaments or abortive stamens.

ĬNÄRCH'ING, the natural union of stems or roots which grow in contact; natural grafting. Also applied in horticulture to a form of grafting in which both stock and scion remain at first attached to their own roots.

ĬNÄRTĬC'ŪLĀTE, not jointed; continuous.

ĬNCȦNĔS'ÇENT, see CANESCENT. Strictly, somewhat or slightly canescent.

ĬNCĀ'NOŬS, see CANESCENT.

ĬNÇĪȘED', irregularly and deeply cut into rather large lobes.

ĬNCLĪNED', gradually bent out of a perpendicular at less than a right angle, as the branches of most deciduous trees.

ĬNCLŪD'ĔD, contained in a cavity and not projecting beyond it; enclosed. Compare EXSERTED.

ĬNCŌMPLĒTE', destitute of some part which is usually present; said especially of flowers which lack one or more of the four sets of primary organs, sepals, stamens, and pistils.

ĬNCŎNSPĬC'ŪOŬS, small in size; not readily observed.

ĬNCRĂS'SĀTE, thickened; especially, gradually thickened or enlarged upward from the base.

ĬNCRĔS'ÇENT, growing.

ĬNCŪBĀ'TIȮN, the period from the time of infection or the sowing of the spores until a bacterium or fungus becomes externally or visibly manifest. Applied mainly to pathogenic bacteria affecting animals to indicate the period from the time the organism enters the body until the disease appears.

IN'CŪBOŬS, having the tip of one leaf overlap the base of the one above it, as in the Jungermanniaceæ. Compare SUCCUBOUS.

ĬNCŬM'BENT, leaning or lying upon; applied to cotyledons when the radicle is folded against the back of one of them (the radicle in such case being dorsal). Compare ACCUMBENT. An anther is *incumbent* when lying on the inside of the filament to which it is attached. Compare VERSATILE.

ĬNCŬR'VĀTE, see INCURVED.

ĬNCŬRVED', bent or curved inward; incurvate.

ĬNDĒÇĬD'ŪOŬS, either evergreen or persistent.

ĬNDĔF'ĬNĬTE, either uncertain or not uniform in number, or too many to be readily counted; numerous; over twenty when applied to stamens. Also applied to objects which have no well-defined boundary or outline.

ĬNDĔF'ĬNĬTE GROWTH, see INDETERMINATE.

ĬNDĔF'ĬNĬTE ĬNFLŌRĔS'ÇENÇE, see INDETERMINATE.

ĬNDĒHĬS'ÇENT, not opening in a definite manner at maturity to discharge the contents. The fruits of the pea and lily are dehiscent, those of the tomato and apple *indehiscent.*

ĬNDĒTĔR'MĬNĀTE, a mode of centripetal inflorescence in which the flowers all arise from axillary buds. Applied also to all stems which do not produce a well-developed terminal bud at the close of the season, as the grape. Compare DETERMINATE.

ĬNDĬF'FĔRENT, undifferentiated; not specialized; as *indifferent* cells or tissues.

ĬNDĬG'ĔNOŬS, strictly native; aboriginal. Compare NATURALIZED.

ĬNDĬVĬD'ŪAL FERTILIZA'TION, a term applied by L. H. Bailey to cross-fertilization between different flowers upon the same plant.

ĬNDŪMĔN'TŬM, any hairy covering upon plants.

ĬNDŪP'LĬCĀTE, having the margins folded inward. Compare INVOLUTE.

ĬNDŪRĂS'ÇENT, becoming hard.

ĬN'DŪRĀTED, hardened.

ĬNDŪ'ŞĬĀTĔD, furnished with an indusium.

ĬNDŪ'ŞĬŬM, an outgrowth of the epidermis covering the sorus in many kinds of ferns; shield. Also applied to a ring of "collecting hairs" below the stigma, as in Lobeliaceæ.

ĬNDŪ'ŞĬŬM, FALSE, see FALSE INDUSIUM.

ĬNDŪ'VĬÆ, any parts of the flower which persist and cover the fruit at maturity; also dead and withered leaves which remain persistent on the stem. Compare RELIQUIÆ.

ĬNĒQUĬLĂT'ĔRAL, unequal sided.

ĬNĔR'MOŬS, unarmed; destitute of spines, prickles, etc.

ĬNFÄRCT'ĀTE, see FARCTATE.

ĬNFĔC'TIOUS. In ordinary use this term has the same sense as Contagious, being applied to all diseases which are communicable from one plant or animal to another by direct contact or otherwise. In a broad sense infectious includes Contagious, as defined under that term, and applies also to diseases originating from germs which are able to vegetate for a time at least outside of the affected plant or animal. In a strict sense Infectious applies only to diseases produced by organisms which have their natural home outside of the infected body.

ĬNFĒ'RĬOR, lower, as an ovary to which the other floral organs are adnate so that they arise from its summit. If the other organs are free from the ovary *they* are inferior and the ovary superior, though the term is seldom applied to them. The *inferior* side of a leaf or flower is the lower or anterior side which faces away from the supporting axis or stem.

ĬNFLĀ'TĔD, puffed up; bladdery.

ĬNFLĔCT'ĔD, see INFLEXED.

ĬNFLĔXED', abruptly bent inward or downward; inflected.

ĬNFLŌRĔS'ÇENÇE, (1) the arrangement of the flowers or flower-clusters on a plant; anthotaxy. Compare PHYLLOTAXY. (2) The portion of the plant which bears the flowers and fruit, i.e., a flower-cluster of any kind.

ĬNFRÅ-ĂX'ĬLLĀRY̆, situated below the axil.

ĬNFRĂC'TĔD, see INFLEXED.

ĬNFRŬCTĔS'ÇENÇE, an inflorescence in fruit; the fruiting portion of a plant, together with its fruit. Little used, and applied mainly to collective fruits.

ĬNFŬNDĬB'ŪLAR, see INFUNDIBULIFORM.

ĬNFŬNDĬB'ŪLĬFÔRM, funnel-shaped; having a tube which gradually enlarges upward and bears a moderately spreading border, as Datura. Compare HYPOCRATERIFORM.

ĬNĬ'TIAL CELLS, the first formed cells of a tissue.

ĬNJEC'TION, filling of intercellular spaces by water, an occurrence which rarely happens.

ĬNNĀTE', (1) said of anthers attached by their base to the apex of the filament; such anthers are sometimes called Vertical; (2) said of an organ or object which grows within the substratum, as "mycelium or perithecia *innate,*" i.e., growing within the tissue of the host.

ĬN'NÊR LĂM'ĬNÅ, the layer of a lignified cell-wall adjoining the inside of the cell. Compare MIDDLE LAMINA.

ĬNNŌVĀ'TION, a new or additional growth or shoot, as the supplementary extensions of the stem in mosses. Applied also to an entire group of offgrowths of the same morphological value if some of the forms are true innovations in their manner of growth; thus Dr. William Trelease applies this term in Epilobium to forms which vary in different species from sessile buds to dense rosettes, running leafy shoots, scaly rhizomes, and filiform bulbiferous stolons.

ĬNŎS'CŪLĀTĬNG, opening into each other; anastomosing.

ĬNSĔRT'ĔD, attached to or growing out of, as stamens *inserted* on the corolla.

ĬNSĔR'TION, the place or mode of attachment.

ĬNSPĬS'SĀTĔD, thickened by drying.

ĬNSTĬP'ŪLĀTE, see EXSTIPULATE.

ĬNTĔGRĬFŌ'LĬOŬS, having entire leaves.

ĬNTĔG'ŪMENT, any covering layer or membrane.

ĬN'TĔR-, in composition, between. Compare ĬNTRA-.

ĬNTĔRĂX'ĬLLĀRY̆, between the axils.

ĬNTĔR'CĂLĀTĔD, inserted between or in the midst of.

ĬNTĔRCÄR'PĔLLĀRY̆, between the carpels.

ĬNTĔRÇĔL'LŪLAR PASS'AGE, a continuous opening between the cells.

ĬNTĔRÇĔL'LŪLAR SPACE, any cavity within the plant. Usually applied to smaller openings than intercellular passages.

ĬNTĔRÇĔL'LŪLAR SUB'STANCE, material extruded from the cells within the plant.

ĬNTĔRÇĔL'LŪLAR SY̆S'TĔM, the intercellular spaces and material of a plant taken together.

ĬNTĔRCŌS'TAL, situated between the ribs of a leaf.

ĬNTĔRFĂSÇĬC'ŪLAR, between the bundles: said of a layer of cambium which extends from one fibrovascular bundle to another.

ĬNTĔRFĪ'LAR, between the filaments, as the resting-spore in the conjugation-tube of Mesocarpus, or the fluid portion of the protoplasm in the hypothetical fibrillar network. Compare ĬNTRAFILAR.

ĬNTĔRFŌLĬĀ'ÇEOŬS, attached to the stem between the bases or petioles of opposite leaves; interpetiolar. Compare ĬNTRAFOLIACEOUS.

ĬNTĔRMĒ'DĬĀTE TĬS'SŪE, all the fundamental tissue in exogens, except that which is immediately associated with the epidermis and the fibrovascular bundles. It includes the pith, medullary rays, and most of the cortex. The term is of little use.

ĬNTĔRMĒ'DĬĀTE ZŌNE, the zone in endogens between the pith and epidermis containing the fibrovascular bundles.

ĬNTĔR'NAL GLAND, a secreting cell, or usually a cluster of secreting cells, within the plant, as those containing essential oil which form the translucent dots in the leaves of the orange.

ĬN'TĔRNŌDE, the portion of a stem between two nodes.

ĬNTĔRPĔT'ĬŌLAR, see ĬNTERFOLIACEOUS.

ĬNTĔRRŬPT'ĔD, said of any surface or series the continuity of which is broken, as a pinnate leaf in which leaflets much larger or smaller than usual are interposed among the others, or a slender stem or root which is contracted at intervals.

ĬNTĔRRŬP'TĔDLY̆ - PĪN'NĀTE, pinnate with small (or sometimes large) leaflets interposed between those of the usual size.

ĬNTĔRSTĬ'TIAL, applied to that method or theory of growth which consists in the interposition of new particles between the older ones instead of additions to the surface.

ĬNTĔX'ĬNE, see ĬNTEXTINE.

ĬNTĔX'TĬNE, a term applied by Fritzsche to the inner part of the extine when, as in Œnothera, it separates as a distinct membrane. Compare EXINTINE.

ĬN'TĬNE, the inner coat of a pollen-grain.

ĬNTŌR'TION, turning to one side from the vertical or any straight line. (Rare.)

ĬN'TRȦ-, in composition, within. Compare INTER-.

ĬNTRȦCÄR'PĔLLĀRY̆, produced inside a carpel.

ĬNTRȦÇĔL'LŪLAR, within a cell.

ĬNTRȦFĂSÇĬC'ŪLAR, within the bundle, as an *intrafascicular* lacuna in Equisetum.

ĬNTRȦFĪ'LAR, within the filament. Compare INTERFILAR.

ĬNTRȦFŌLĬĀ'ÇEOŬS, within a leaf; between the leaf and stem, as the stipules of Polygonum; intrapetiolar. Compare INTERFOLIACEOUS.

ĬNTRȦLĂM'ĔLLAR, within special layers or lamellæ, as the trama of Hymenomycetes.

ĬNTRȦMÄR'ĠĬNAL, situated within but near the margin.

ĬNTRȦMĂT'RĬCAL, in a matrix or nidus.

ĬNTRȦPĔT'ĬŌLAR, (1) inside or beneath the petiole, as the buds of sumach and sycamore (subpetiolar); (2) between the petiole and the stem, as the stipules of sycamore and most buds; intrafoliaceous.

ĬNTRȦVĂĠ'ĬNAL, within the sheath: applied to branches in grasses which in their growth do not break through the base of the sheath of the subtending leaf, but push upward between the sheath and the stem. Compare EXTRAVAGINAL.

ĬNTRŌDŪÇED', applied to plants brought from another country, and growing spontaneously unless otherwise noted. Compare NATURALIZED.

ĬNTRŌFLĔXED', bent strongly inward. Between Incurved and Infracted.

ĬNTRÔRSE', facing or turning inward. Applied to anthers which open on the side next the pistil.

ĬNTRṲD'ĔD, appearing as if pushed inward or indented; intruse.

ĬNTŬSSŬSÇĔP'TION, the intercalation or formation of other particles among those already present.

ĬN'ŪLĬN, a material isomeric with (resembling) starch which replaces that substance in many Compositæ.

ĬNVĂĠ'ĬNĀTĔD, inclosed in a sheath.

ĬNVÊRSE', see INVERTED.

ĬNVÊRT'ĔD, having a position or mode of attachment the reverse of that which is usual; inverse. A seed or ovule is *inverted* when attached to the top of the ovary. It is then, however, more properly called "suspended." Compare ERECT and ASCENDING.

ĬNVÊRT'ĔD SŪPĔRPŌṢĬ'TION, the situation of accessory buds below the principal bud or one first formed. Compare DIRECT SUPERPOSITION.

ĬNVŎL'ŪÇĔL, a partial or secondary involucre, as one subtending a partial umbel.

ĬNVŌLŪÇĔL'LŬM, see INVOLUCEL.

ĬNVŌLŪ'CRAL, pertaining to an involucre.

ĬNVŌLŪ'CRĀTE, having an involucre; involucred.

ĬN'VŌLŪCRE, a set of bracts immediately subtending a flower or inflorescence; involucrum.

ĬN'VŌLŪCRED, see INVOLUCRATE.

ĬNVŌLŪ'CRĔT, see INVOLUCEL.

ĬNVŌLŪ'CRŬM, see INVOLUCRE.

ĬN'VŌLŪTE, rolled inward from both sides. Compare CONVOLUTE.

ĬNVŌLŪ'TION-FŌRM, a swollen bladder-like form in Schizomycetes, supposed to be a diseased condition of the form with which it is found associated. (De Bary.)

ĬNVŌLŪ'TION-PĒ'RĬŎD, see RESTING-PERIOD.

ĬNVŌLŪ'TION-STĀGE, see RESTING-STAGE.

ĬRRĔG'ŪLAR, denoting flowers in which one or more of the organs of a set are different in size or form from the others. Irregularity occurs most frequently in the petals. The pea is an example of an irregular flower, the tulip of a regular. Compare SYMMETRICAL.

ĬRRĔG'ŪLAR PĒLŌ'RĬA, a teratological condition in which an irregular flower becomes regular by the formation of the irregular parts in increased number. Compare REGULAR PELORIA.

ĬRRĬTĀBĬL'ĬTY̆, having the power of movement in a definite manner under the influence of external stimuli, as in the coiling of tendrils, twining of stems, or "sleep" of leaves; sensitiveness. Compare EXCITABILITY and CONTRACTILITY.

ĪSĀDĔL'PHOŬS, having an equal number of stamens in each adelphia.

ĪSŌBĪLĂT'ĔRAL, having two sides alike in form and structure, as the leaves in the iris.

ĪSŎB'RĬOŬS, applied to the embryo of dicotyledons because both are equally developed; isodynamous. Little used.

ĪSŌCHRŌ'ŬS, uniform in color throughout; unicolor. Compare CONCOLOR.

ĪSŌDY̆'NĀMOŬS, equally developed.

ĪSŎG'ĀMY̆, the conjugation of gametes of similar form. Compare OÖGAMY.

ĪSŎG'Y̆NOŬS, having the pistils of a flower all alike. Compare HETEROGYNOUS.

ĪSŌMĔR'ĬC, see ISOMEROUS.

ĪSŎM'ĔROŬS, having the same number of organs in each floral whorl; isomeric. There may be more than one whorl of any of the kinds of organs. Compare HETEROMEROUS.

ĪSŎPH'ŌROŬS, transformable into something else, as "Actinia is an *isophorous* form of Dendrobium."

Ī'SŌSPŌRE, said of a spore when all are alike, as in ferns. (Rare.) Compare MACROSPORE and MICROSPORE.

ĪSŎS'PŌROŬS, not having macrospores and microspores; homosporous. Compare HETEROSPOROUS.

ĪSŌSTĔM'ŌNOŬS, having the stamens equal in number to the petals. More properly, having the stamens and petals each in one whorl and of equal number. Compare ANISOSTEMONOUS, MEIOSTEMONOUS, DIPLOSTEMONOUS, and OBDIPLOSTEMONOUS.

ĪSŎS'TŌMOŬS, having calyx and corolla of equal size. (Rare.)

ĬSTH'MŬS, the constricted portion between the two half-cells in most desmids.

JOINT, a node; the septum between two cells in a filament; articulation.

JŪ'BÀ, a loose panicle. (Obs.)

JŪ'GŬM (pl. **Jū'gà**), (1) one of the ridges on an umbelliferous fruit; (2) a pair of leaflets in a pinnate leaf.

JŪLĀ'ÇEOŬS, see AMENTACEOUS.

JŪ'LĬFÔRM, resembling an ament or catkin.

JŪ'LŬS, see AMENT.

JŪVĒNĔS'ÇENCE, see REJUVENESCENCE.

KĂRŸŌKĬNĒ'SĬS (also spelled Caryocinesis), Schleicher's term, which has been generally adopted, for the transformations of the nucleus during cell-division; indirect division of Fleming.

KĂRŸŎL'ŸSĬS, the dissolution of the nucleus or some part of it.

KĂRŸŌMĪTŌ'SĬS, see under MITOSIS.

KĂR'ŸŌPLĂŞM, see NUCLEOPLASM.

KĂRŸŌSŌ'MÀ (pl. **Kărÿōsŏm'ătà**), a consolidated mass of microsomata in a nucleus.

KĂTÀBŎL'ĬC, a term applied by Geddes to disruptive, destructive, or descending metabolism, accompanying the waste of tissues, resulting in the formation of chemical prodncts of simpler composition; catabolic. Compare ANABOLIC.

KÀTHŌD'ĬC, see CATHODIC.

KEEL, a ridge somewhat resembling the keel of a boat; particularly the two inferior petals of a papilionaceous flower which are more or less united into a keel-shaped body; carina.

KEELED, having a keel or longitudinal ridge; carinated.

KĔRÀMĬD'ĬŬM, see CERAMIDIUM.

KEY, see SAMARA.

KEY-FRUIT, see SAMARA.

KID'NĔY-FŌRM, see RENIFORM.

KID'NEY-SHAPED, see RENIFORM.

KNEE, a kind of knot which projects upward into the air from the roots of the bald cypress (*Taxodium distichum*) and some other trees. Produced mainly in wet soil, and formerly supposed to serve for aëration, but now believed to be an organ of strength. See PNEUMATODE. The term *knee* is also applied to any abruptly bent or knee-shaped organ.

KNEE-JOINTED, see GENICULATE.

KNEEPAN-SHAPED, see PATELLIFORM.

KNOT, a node or swollen joint; a protuberance on the surface of a tree, as where a branch has been removed and the cut or broken surface is more or less overgrown; a place in the wood where the tissues are displaced by an injury or by the passage of a branch.

KNOTTED, cylindrical, and swollen at intervals, somewhat like a knotted cord.

LÀBĔL'LŬM, the large lower petal of an orchid; lip.

LĀ'BĬĀTE, gamopetalous, with two divisions, anterior and posterior; two-lipped; bilabiate. The two lips of a labiate flower are usually unequal and the flower irregular, as in snap-dragon.

LĀ'BĬŌSE, having the petals of a polypetalous corolla arranged so as to imitate the labiate form. (Rare.)

LĀ'BĬŬM, the lower lip of a labiate flower.

LĂÇ'ĔRĀTE, having the margin deeply cut into irregular segments as if torn. Compare EROSE, ĬNCISED, LACINIATE.

LĂÇ'ĔRĀTED, see LACERATE.

LĂCH'RȲMÆFÔRM, see TEAR-SHAPED.

LȦCĬN'ĬȦ (pl. Lȧcĭn'iǣ), a segment of a laciniate leaf.

LȦÇĬN'ĬĀTE, deeply cut into narrow incisions, more irregular and larger than Fimbriate; slashed.

LȦÇĬN'ĬFÔRM, fringe-like.

LȦÇĬN'ŪLĀTE, finely laciniate. Compare LACINULOSE.

LȦÇĬN'ŪLŌSE, lacinulate or bearing little fringes.

LĂCTĔS'ÇENT, resembling or producing milk or latex.

LĂCTĬF'ĔROŬS, producing or conveying latex; laticiferous.

LĂCTĬF'ĔROŬS VĔS'SĔLS, see LATICIFEROUS VESSELS.

LȦCŬ'NȦ (pl. Lȧcū'nǣ), a large deep depression on the surface, or open space between the cells.

LȦCŪ'NAR, having, resembling, or pertaining to lacunæ.

LĂC'ŪNŌSE, (1) perforated with rather large holes. Compare FORAMINATED. (2) Having depressions in the surface larger, more irregular, or more irregularly placed than in Alveolate.

LĂC'ŪNŌSE - RUGŌSE', marked with deep broad irregular wrinkles, as the shell of the walnut or pit of the peach. Compare RUMINATED.

LȦCŬS'TRĪNE, growing in or on the margins of lakes.

LADDER-CELLS, see SCALARIFORM VESSELS.

LÆV'ĬGĀTE, smoothed, as if polished; lævis.

LÆ'VĬS, smooth; having an even surface devoid of hair or roughness of any kind; opposed especially to asperate, striate, sulcate, or any unevenness. Compare NITID and GLABROUS.

LȦĠĒ'NĬFÔRM, shaped like a Florence flask (the ordinary bulbous flask of chemists). The term bottle-shaped means the same, or nearly the same.

LĂM'EL, see LAMELLA.

LȦMĔL'LȦ (pl. Lȧmĕl'lǣ), diminutive of Lamina; a thin plate, as one of the "gills" beneath the cap of a mushroom.

LĂM'ĔLLĀTE, composed of thin plates or scales; lamellose.

LȦMĔL'LĬFÔRM, in the form of a plate or scale.

LĂM'ĔLLŌSE, see LAMELLATE.

LĂM'ĬNȦ (pl Lăm'ĭnǣ), the blade of a leaf or limb of a petal or sepal.

LĂM'ĬNĀTĔD, consisting of plates, scales, or layers; plated.

LĂM'ĬNĀTĔD BŬLB, see TUNICATED BULB.

LĀ'NĀTE, covered with long curled hairs like wool; lanose; lanuginous; woolly.

LĂN'ÇĒŌLĀTE, tapering abruptly toward the base and gradually toward the apex, like the head of a lance.

LANCE-OVATE, between lanceolate and ovate, but approaching nearer the latter.

LĀ'NŌSE, see LANATE.

LȦNŬ'ĠĬNOŬS, see LANATE.

LĂP'ĬDŌSE, growing in stony places.

LĂPPĀ'ÇEOŬS, bur-like.

LÄR'VAL STATE, resting state, as the sphacelium of ergot. (W. G. Smith.) Seldom used.

LĂSĬŌCĂR'POŬS, having pubescent fruit.

LĀ'TENT, remaining dormant beyond the usual time, or until called into growth by some particular stimulus.

LĀ'TENT BŬD, see DORMANT BUD and ADVENTITIOUS BUD.

LĀ'TENT PĒ'RĬŌD, see DORMANT STATE and RESTING PERIOD.

LĂT'ĔRAL, attached to the side of an organ.

LĂT'ĔRAL, n., a side branch or root.

LĂT'ĔRAL BŬD, one situated on the side of a branch, usually in the axil of a leaf. Compare TERMINAL BUD.

LĂT'ĔRAL DĒHĬS'ÇENÇE, when the opening of an anther takes place upon one or both sides, instead of on the surface facing or opposite to the pistil.

LĂT'ĔRAL NŪCLĒ'ŌLŬS, see PARANUCLEOLUS.

LĂT'ĔRAL NŬ'CLĔŬS, see PARANUCLEUS.

LĂT'ĔRAL PLĀNE, of a flower or other lateral structure, a plane passing from side to side at right angles to the median plane. The lateral plane of a leaf, for example, would pass between its upper and lower surfaces.

LĀ'TĔX, a viscid milky fluid found in certain plants, as the lettuce. See LATICIFEROUS VESSELS.

LĀ'TĔX TUBES, see LATICIFEROUS VESSELS.

LĂTĬÇĬF'ĔROŬS VĔS'SĔLŞ, special anastomosing tubes containing the latex in such plants as have milky juice; latex tubes.

LĂTĬFŌ'LĬĀTE, having broad leaves. Compare STENOPHYLLOUS and ANGUSTIFOLIATE.

LĂTĬFŌ'LĬOŬS, see LATIFOLIATE.

LĂTĬSĔP'TĀTE, having a broad septum or partition;—applied to pericarps. Compare ANGUSTISEPTATE.

LĂT'TĬÇED, see CLATHRATE.

LĂT'TĬÇED CELL, see SIEVE-TUBE.

LĂX, loose; the opposite of close or crowded. An inflorescence may be *lax* because its pedicels are flaccid or slender.

LĔAD'ĔR, the upper portion of the primary stem of a tree, or a central upright branch which extends beyond the rest of the head.

LĔAF, an appendage to the stem, definite in position, and usually expanded to receive air and light. Besides leaves of the usual forms which serve as foliage there are other forms for additional purposes, including those which form the various parts of a flower. See PHYLLOME.

LĔAF-ĂRRĀNĠE'MENT, see PHYLLOTAXIS.

LĔAF-BLĀDE, the expanded portion of ordinary leaves; lamina.

LĔAF-BŬD, a bud which immediately upon expansion produces leaves only. Compare FLOWER-BUD.

LĔAF-ÇȲ'CLE, the course of a spiral from the insertion of a leaf through that of intermediate ones to the insertion of the next leaf directly above or below the place of starting.

LĔAF-GREEN, see CHLOROPHYLL.

LĔAF'LĔT, one of the divisions or blades of a compound leaf. Leaflets are often articulated to the common petiole, and in order to be distinguished from segments they must themselves have distinct petioles, however short.

LĔAF-SCÄR, the place on a stem from which a leaf has fallen.

LĔAF-SHĔATH, see SHEATH.

LĔAF-STALK, see PETIOLE.

LĒAF-TĔN'DRĬL, a tendril which is a transformed leaf or part of a leaf.

LĔAF-TRĀÇE, a fibrovascular bundle, or collection of bundles, while on its passage from the fibrovascular system of the stem to the leaf.

LĔATH'ĔRY̆, see CORIACEOUS.

LĔFT. The left margin of a leaf, petal, or other foliar organ is the one upon the left hand as its stem is held downward or toward the observer and the organ is viewed upon the upper or inner surface. The left side of a lateral flower is the side upon the left hand as the observer stands in front with the stem erect and the flower between him and the main axis. See remarks under Dextrorse.

LĔĠĬT'ĬMĀTE FĔRTĬLĬZĀ'-TION, fertizliation in the usual or natural manner, either close or cross fertilization, whichever in the particular case occurs in nature.

LĔG'ŪME, the fruit of the Leguminosæ. It is monocarpellary, with the seeds attached to the ventral side, and usually monolocular and dehiscent into two valves by ventral and dorsal sutures. The pea and bean are examples. Compare LOMENT.

LĒGŪ'MĬNOŬS, bearing legumes, or relating to plants of the order Leguminosæ.

LEIO-, a prefix from the Greek, meaning smooth.

LĔN'TĬÇĔLS, small developments of cork at particular points on the surface of many kinds of stems. They appear during the first season's growth, and in some cases, at least, originate beneath the stomata. They probably aid in the transfer of gases and aqeuous vapor.

LĔNTĬÇĔLLE', see LENTICEL.

LĔNTĬC'ŪLÀ (pl. **Lĕntĭc'ūlæ**), see LENTICEL.

LĔNTĬC'ŪLAR, in the form of a double convex lens; lentiform.

LĔN'TĬFÔRM, see LENTICULAR.

LĔNTĬĠ'ĬNŌSE, see LENTIGINOUS.

LĔNTĬĠ'ĬNOŬS, covered with small dots, like dust, scurf, or freckles; lentiginose.

LĔP'AL, a rudimentary sterile stamen, or organ supposed to be such, usually in the form of a nectar-gland or petaloid scale; gland of the torus. See STAMINODIUM.

LĒPĂN'THĬŬM, an old term for a petal which contains a nectary.

LĔP'ĬDĒS, epidermal scales of various kinds, especially when attached at the centre. Compare RAMENTA.

LĔP'ĬDŌTE, scaly; scurfy; leprous; leprose.

LĔP'RŌSE, see LEPIDOTE.

LĔP'ROŬS, see LEPIDOTE.

LĔP'TÀ-, or **LĔP'TŌ-,** a prefix from the Greek, meaning small, thin, or slender.

LĔP'TŌME, see PHLOËM. Applied by Potonié to the phloëm-like portion of the fibrovascular bundle in vascular cryptogams. Compare HADROME.

LĔPTŌPHLŌ'ĔM, a term applied by Vaizey to a portion of the central fascicle or strand in the seta of Polytrichaceæ (in mosses) which he considers to be rudimentary phloëm.

LĔPTŌPHȲL'LOŬS, having scale-like leaves.

LĔPTŌSPŌRĂN'ĠĬŬM, a sporangium derived from a single epidermal cell, as in the true ferns, in distinction from one derived from a group of epidermal cells (eusporangium), as in the Ophioglossaceæ and Marattiaceæ.

LĔPTŌXȲ'LĔM, so-called rudimentary xylem in the seta of Polytrichaceæ. (Vaizey.)

LEU-, or **LEUCO-,** a prefix from the Greek, meaning white.

LEŬCĂN'THOŬS, white-flowered.

LEŬ'ÇĪTE, see LEUCOPLAST.

LEŬ'CŌPLĂST, a colorless protoplasmic granule (plastid) such as those in which starch originates; starch-builder; leucoplastid; amidoplast; anaplast; leucite proper. See PLASTID.

LĔV'ĬGĀTE, see LÆVIGATE.

LĒ'VĬS, see LÆVIS.

LĪ'ĀNE, a woody climbing or twining plant of considerable size; applied only to those which abound in some tropical forests.

LĪ'BĔR, the inner layer of bark, consisting chiefly of the phloëm of the fibrovascular system and therefore containing the bast-tissue; bast.

LĪ'BRĬFÔRM CELL, a narrow thick-walled cell of woody tissue resembling bast; wood-fibre.

LID, see OPERCULUM.

LID-CELLS, terminal cells of the neck in some archegonia which for a time close the canal; stigmatic cells.

LIFE-ÇȲ'CLE, the course of development from any given stage, as the spore or seed, back to the same stage again.

LĬG'NĒOŬS, woody, or pertaining to wood.

LĬGNĬFĬCĀ'TION, the process of becoming converted into wood; sclerosis.

LĬG'NĪNE, a secondary deposit, forming the greater part of the bulk of ordinary wood. It contains less oxygen than cellulose, and according to Payen has the formula $C_{35}H_{24}O_{20}$. It is also written Lignin, and is the same as Sclerogen.

LĬG'NŬM, see WOOD.

LĬG'ŪLĂ (pl. **Lig'ūlæ**), see LIGULE.

LĬG'ŪLĀTE, (1) having a ligule; (2) strap-shaped, i.e., linear and about four to six times as long as broad.

LĬG'ŪLĀTE FLŌ'RĔT, one of the marginal flowers of a head in Compositæ, bearing a ligulate or strap-shaped corolla; ray-floret. Compare TUBULAR FLORET.

LĬG'ŪLE, (1) a strap-shaped corolla in Compositæ, such as those on the outer margin of the head in most sun-flowers; (2) a membranous appendage on the inner side of the leaf in many grasses and some other endogens at the top of the sheath.

LĬGŪLĬFLŌ'ROŬS, said of a head of flowers in Compositæ which contains ligulate flowers only, as in Chicory.

LĬMB, the expanded portion of a petal or sepal. Compare BLADE and BORDER.

LĬM'BĀTE, see BORDERED.

LĬM'BŬS, border, blade, or limb. (Obs.)

LĬM'ĬTĬNG CELL, see HETEROCYST.

LINE, one twelfth of an inch.

LĬN'ĔAR, very narrow, with the margins parallel or nearly so.

LĬN'ĔĀTE, marked with fine parallel lines, especially of color. Compare STRIATE and ACICULATED.

LĬN'ĔŌLĀTE, diminutive of Lineate.

LĬN'GUÆFÔRM, see TONGUE-SHAPED.

LĬN'GUĬFÔRM, see TONGUE-SHAPED.

LĬN'GŪLĀTE, see TONGUE-SHAPED.

LĪ'NĪNE, a term proposed by Schwarz and adopted by Strasburger (1888) for the substance of the hyaloplasmic filaments of the nucleus in a state of repose.

LĬNNÆ'AN SYS'TEM, the system of classification devised by Linnæus, founded upon the number and arrangement of the stamens and pistils; sexual system.

LIP, (1) one of the divisions (especially the lower) of a labiate calyx or corolla; (2) the labellum in orchids.

LĪPŎX'ĔNOŬS, said of a parasite which leaves its host and completes its development independently at the expense of the reserve material appropriated from the host. (De Bary.) Compare METŒCIOUS.

LĬPPED, see LABIATE. Applied in pomology to the cavity of an apple when a broad projection of the flesh protrudes against the stem. Compare FOLDED.

LĬRĔL'LĂ (pl. **Lĭrel'læ**), a linear sessile apothecium in lichens.

LĬTHŎPH'ĬLOŬS, see SAXICOLOUS.

LĬTH'ŌCÄRP, see CARPOLITE.

LĬTH'ŌÇŸST, a cell containing a cystolith.

LĬTHŌSPĔR'MOŬS, having very hard seeds.

LĬT'TŌRAL, growing upon shores, especially upon the sea-shore between tide-marks. Compare MARINE.

LĬV'ĬD, lead-color; bluish brown or gray.

LŌ'BĀTE, lobed; said of a leaf the margin of which is divided about to the middle into rounded parts with broad sinuses.

LŌBE, a rounded portion of a leaf or petal; any division of a leaf larger than a tooth which is not a leaflet.

LŌBED, see LOBATE.

LŎB'ŪLĀTE, having small lobes.

LŎB'ŪLE, a small lobe.

LŌÇĔL'LĀTE, having small or secondary cavities (locelli).

LŌÇĔL'LŬS (pl. **Lōçĕl'lī**), diminutive of Loculus; a little cavity. Used by some in the sense of Loculus for a compartment in an ovary or anther.

LŎC'ŪLĂMENT, see LOCULUS.

LŎC'ŪLAR, having a cavity or cavities; loculose.

LŎCŪLĬCĪ'DAL, dehiscent throughout the middle of the back of each cell at a dorsal suture. Compare SEPTICIDAL.

LŎC'ŪLŌSE, locular; cellular; especially, having numerous cavities, as the pith of poke, *Phytolacca decandra.*

LŎC'ŪLOŬS, see LOCULAR.

LŎC'ŪLŬS (pl. **Lŏc'ūlī**), the cell of an ovary, anther, or similar cavity.

LŌCŬS'TȦ, see SPIKELET.

LŌDĬC'ŪLȦ (pl. **Lōdic'ūlæ**), see LODICULE.

LŎD'ĬCŪLE, one of the small scales at the base of the grain between the stamens and palet of many grasses.

LŌ'MENT, a jointed or plurilocular legume formed by spurious transverse dissepiments, as in Desmodium. Some loments are articulated and separable at the septa or pseudo-septa between the seeds.

LŌMĔNTĀ'ÇEOŬS, bearing or resembling a loment.

LŌMĔN'TŬM, see LOMENT.

LŎNǴĬTŪ'DĬNAL SȲS'TĔM, an obsolete term for fibrovascular system.

LŌ'RĀTE, having the form of a thong or strap; ligulate; linear. Compare FLABELLIFORM.

LŪ'BRĬCOŬS, slippery.

LŪ'ÇĬD, clear. Often used in the sense of Nitid.

LŬM'BRĬCAL, about the shape of an earth-worm. Compare VERMICULAR.

LŪ'MĔN (pl. **Lū'mĭnȧ**), the cavity of a cell or filament, or any part or spot which seems to emit light.

LŪ'NAR, (1) pertaining to the moon; (2) in the form of a half-moon or crescent, as a *lunar* spot upon a leaf; semilunar; semilunate. Compare LUNATE.

LŪ'NĀTE, said of an organ having the form of a half-moon or crescent, or which is marked with one or more spots of that form, as a *lunate* leaf. Compare LUNAR.

LŪ'NĀTĔD, having lunar markings; lunate.

LŪ'NĬFÔRM, half-moon or crescent shaped; lunate.

LŪ'NŪLĀTE, diminutive of Lunate.

LŪ'PŪLĬNE, a yellow waxy powder resembling pollen, upon the scales of the hop, containing the active principle, which is bitter, aromatic, and tonic.

LŪPŪLĪ'NOŬS, resembling a head or cone of hops.

LŪ'RĬD, dingy brown.

LŪTĔS'ÇENT, yellowish.

LŪ'TĔŬS, bright orange yellow. Compare FLAVUS.

LȲCŎT'RŌPAL, see LYCOTROPOUS.

LȲCŎT'RŌPOŬS, said of an otherwise orthotropous ovule which is curved like a horseshoe.

LȲMPHĂT'ĬC, clear, like water; not clouded: said, for example, of ordinary plant-hairs containing clear cell-fluid in distinction from those which are glandular, in which the cell-contents are usually granular and turbid.

LȲ'RĀTE, pinnatifid, with the lobes decreasing in size toward the base.

LȲ'RĀTELȲ PIN'NATE, pinnate, with the pinnæ decreasing in size toward the base of the leaf.

LȲRE-SHAPED, see LYRATE.

LȲSĬǴĔNĔT'ĬC, see LYSIGENIC.

LȲSĬǴĔN'ĬC, formed by the disorganization of cells; applied to the formation of certain intercellular spaces which originate by the breaking down of contiguous cells; lysigenetic;

lysigenous. Compare SCHIZOGENIC.

LŸSĬG'ĒNOŬS, see LYSIGENIC.

MĂCRĂN'DROŬS, having large or elongated male plants in algæ.

MACRO-, large or long.

MĂCRŌÇĔPH'ÀLOŬS, a term sometimes applied to embryos which have the cotyledons very large and consolidated, as in horse-chestnut. Compare MACROPODOUS.

MĂCRŌGŌNĬD'ĬŬM (pl. **Măcrōgōnĭd'ĭà**), see MACROSPORE.

MÀCRŎPH'ŸLLOŬS, having long or large leaves.

MÀCRŎP'ŌDOŬS, a term of little importance applied to a leaf having a long petiole, or to an embryo, chiefly in monocotyledons, having a large radicle.

MĂCRŌSPŌRĂN'ĠĬŬM, a sporangium containing macrospores; megasporangium; oösporangium.

MĂC'RŌSPŌRE, a female or archegonium-bearing spore in some vascular cryptogams. The macrospores are larger than the male or microspores.

MĂCRŌSPŌ'RŌPHŸL, the leaf bearing the macrosporangium in the heterosporous Pteridophyta. It is the homologue of the carpel in flowering plants.

MĂCRŌSTŸ'LOŬS, applied to heterostylous flowers with long styles and short filaments. See remark under Microstylous.

MĂC'ŪLĀTE, spotted or blotched; maculose.

MĂCŪLĀ'TION, the pattern or arrangement of the spots upon a plant.

MĂC'ŪLŌSE, see MACULATE.

MAIDEN (Hort.), a term sometimes applied to a tree or other plant of one year's growth from the bud or graft. Also applied to any plant which has not fruited, and in forestry to a seedling in distinction from a tree which has sprung from an old root or stub.

MĂL'ÀCOID, mucilaginous.

MALE, bearing stamens only; staminate; bearing only male organs of whatever kind, or pertaining to individuals which bear them.

MALE FLOWER, staminate flower.

MĂLPĬGHĬĀ'ÇEOŬS HAIRS, stellate hairs attached by their middle, as in the order Malpighiaceæ.

MĂLPĬG'HĬAN ÇĔLLȘ, palisade-like cells in which one or more light lines are present, found in the seed-coats of Leguminosæ, Malvaceæ, etc.

MĂM'MĬFÔRM, breast-shaped.

MÀMMĬL'LÀ (pl. **Màmmĭl'læ**), a small nipple or teat-like prominence, as on some pollen-grains. More short and obtuse than Papilla. The term is sometimes applied to the apex of the nucleus of an ovule.

MĂM'MĬLLĀTE, nipple-shaped, or bearing mammillæ. Compare PAPILLATE.

MĂM'MĬLLĂTĔD, bearing mammillæ.

MĂM'MŌSE, breast-shaped, or bearing breast-shaped prominences.

MĂN'ĬCĀTE, having a coating of entangled hairs which may be removed entire. Compare FLOCCOSE and PANNOSE.

MÀNŪ'BRĬŬM (pl. **Mănū'brĭà**), a cylindrical cell which projects inward from the centre of each of the eight shields composing the wall of the globule in Characeæ. The manubria support the apparatus which contains the antherozoids.

MÄR'BLED, covered with faint irregular wide stripes, or traversed by vein-like markings like those which occur in certain kinds of marble; marmorate.

MÄRÇĔS'ÇENT, withering but not falling off, as the corolla of lobelia.

MÄR'ĠĬNAL VEIL, in Hymenomycetes (mushrooms, etc.), a membrane stretching from the margin of the pileus to the surface of the stipe in the young sporophore and covering the hymenium; velum partiale.

MÄR'ĠĬNĀTE, having a margin differing in texture, form, or color from the remainder of the surface; margined; edged.

MÄR'ĠĬNED, see MARGINATE.

MȦRĬNE', growing in the sea or upon the immediate sea-shore. Compare LITTORAL and MARITIME.

MĂR'ĬTĬME, growing upon or near the sea-shore.

MÄR'MŌRĀTE, see MARBLED.

MȦSKED, see PERSONATE.

MĂS'SŪLȦ (pl. **Măs'sūlæ**), one of the lumps of mucilage derived from the disorganized tapetum of the microsporangium of Azolla and enclosing a number of spores. Also applied to an adherent group of pollen-grains derived from a single mother-cell, as the ultimate groups in a pollen-mass in orchids, each containing four grains.

MȦST, a popular name for the fruit of the oak, chestnut, beech, and some other forest trees.

MĂS'TŌID, large teat-shaped. (Rare.)

MĀ'TRĬX, the place or body upon or within which anything grows.

MĂTTŬL'LA, the fibrous material at the base of the petioles of palms; reticulum.

MĂTŪRĔS'ÇENT, approaching maturity.

MĂTŪTĪ'NAL, pertaining to the morning; expanding in the morning, as the flowers of morning-glory.

MĒAL'Ȳ, see FARINACEOUS.

MĒ'DĬAL, pertaining to the middle; median.

MĒ'DĬAN, see MEDIAL.

MĒ'DĬAN LINE, a line along the centre or axis of a bilateral organ, as along the midrib of a leaf.

MĒ'DĬAN PLANE of a flower or other lateral structure, the same as antero-posterior plane, i.e., a vertical plane which bisects an organ into right and left halves. The median plane of a leaf would pass through the midrib above and below. Compare LATERAL PLANE.

MĒDŬL'LȦ, pith; also central tissue of some other kinds, as of the thallus in lichens.

MĔD'ŬLLĀRȲ, pertaining to medulla or pith.

MĔD'ŬLLĀRȲ RAYS, vertical plates of cellular tissue in the fibrovascular system of exogens connecting the pith and the cortex. They form the glistening spots called "silver grain" which are seen on the surface of wood which is split radially. Only the rays formed the first season extend entirely to the pith. See XYLEM RAY and PHLOËM RAY.

MĔD'ŬLLĀRY̆ SHEATH, the ring of primary xylem bundles which project into the pith. Compare CORTICAL SHEATH.

MĔD'ŬLLĀRY̆ SYS'TEM, a term sometimes applied to the whole fundamental system, but usually, and more properly, employed to include only the pith and medullary rays, or fundamental tissue within the cortex.

MĔD'ŬLLŌSE, pith-like.

MĔGȦLŌGŌNĬD'ĬŬM, see MACROSPORE.

MĔGȦRHȲ'ZOŬS, large-rooted.

MĔGȦSPŌRĂN'GĬŬM, see MACROSPORANGIUM.

MĔG'ȦSPŌRE, see MACROSPORE.

MEĪ'ŌPHY̆LLY̆, the suppression of one or more leaves or foliar organs in a whorl; miophylly. It differs from Abortion in the fact that the suppressed organs have never started to grow.

MEĪŌSTĔM'ŌNOŬS, having fewer stamens than petals. This is its usual meaning. Compare ISOSTEMONOUS.

MEĪ'ŌTĂXY̆, the complete suppression of a whole set of organs, as all of the corolla or all of the stamens.

MĔLȦNŌSPÊRM'OŬS, having dark-colored seeds or spores.

MĔLLĬF'ĔROŬS, honey-bearing.

MĔL'ȮN-SHĀPED, oval, with depressed lines running from end to end, as in most muskmelons.

MĔLT'ĬNG, easily dissolving in the mouth, or under slight pressure. A descriptive character in certain varieties of pears.

MĔM'BĔR, a term employed to designate any part of a plant when treated with reference to its position and fundamental structure, but not with reference to function. Thus, all leaves are one and the same *member*, though they may be developed in different ways to form different organs. Compare ORGAN.

MĔMBRȦNĀ'ÇEOŬS, see MEMBRANOUS.

MĔM'BRĀNE, a very thin expanded tissue, serving the purpose of separation or protection.

MĔMBRĀ'NĒOŬS, see MEMBRANOUS.

MĔM'BRȦNOŬS, thin, soft, and usually translucent, like a membrane. Compare SCARIOUS.

MĔNĬS'CO̅ID, shaped like a meniscus or concavo-convex lens.

MĔN'TŬM, a forward projection of the foot of the column in some orchids.

MĔR'ĬCÄRP, one of the carpids or ripened carpels of a schizocarp, especially one of the halves of the cremocarp in Umbelliferæ; diachænium; hemicarp.

MĔR'ĬDĬSK, a name proposed by Clos for any process upon the receptacle, aside from the floral organs, whether glandular or not. See LEPAL.

MĔRĬSMĂT'ĬC, dividing into parts by the formation of septa.

MĔR'ĬSPŌRE, one of the cells of a compound spore.

MĔR'ĬSTĔM, tissue in an early condition or nascent state while the cells are still undergoing cell-division, as in cambium; formative tissue; generating tissue.

MĔR'ĬSTĔM, PRIMARY, see PRIMARY MERISTEM.

MĔR'ĬSTĔM, SECONDARY, see SECONDARY MERISTEM.

MĔRĬSTĔMĂT'ĬC, consisting of meristem.

MĔR'ĬTHĂLL, see INTERNODE.

MĔRĬTHĂL'LŬS (obs.), see INTERNODE.

MĔRŌBLĂS'TĬC, applied to the formation of the embryo from a part only of the oösphere, as in some Gymnosperms.

MĔS'ŌBLĂST (obs.), see NUCLEUS.

MĔS'ŌCĂRP, the middle layer of a pericarp when distinguishable into three layers. This layer is sometimes developed as a sarcocarp. Applied mainly to stone-fruits.

MĔS'ŌCHĬL, the central part of the labellum in such orchids as have that organ separated into three distinct portions.

MĔSŌPHLŒ'ŬM, the middle or green layer of the bark, between the liber and the outer (usually dry) epiphlœum or corky layer; cellular-envelope.

MĔS'ŌPHЎLL, all the fundamental tissue of a leaf within the epidermis.

MĔSŌPHЎL'LŬM, see MESOPHYLL.

MĔSŎPH'ЎTŬM (obs.), see COLLAR. Also applied by Lindley to the line of demarcation between internode and petiole.

MĔS'ŌSPĔRM, see SECUNDINE.

MĔS'ŌSPŌRE, the middle coat of a spore when there are three.

MĔSŌSTЎ'LOŬS, having styles of intermediate length in heterostyled plants.

MĔSŌTHĒ'ÇĬŬM, the central of the three layers of cells in the coat of an immature anther; fibrous layer. The cells of this layer have annular thickenings. In the mature anther the proper endothecium has often disappeared, leaving the mesothecium as the inner layer, which is then usually called endothecium.

MĔS'TŌME, a term applied by Schwendener to fibrovascular tissue on functional grounds, in distinction from cortical tissue, which (serving for support chiefly) he termed stereome. Strictly, however, the term mestome refers only to the essential or conductive portion of a bundle, the bundle-sheath being strengthening tissue and therefore stereome.

MĒTĂB'ŌLĬSM, the chemical changes which take place in a plant in the process of growth; metastasis; transmutation. It includes both Anabolism and Catabolism.

MĔTȦĠĔN'ĔSĬS, the production of sexual individuals by nonsexual means, either directly or through intervening sexless generations; alternation of generations; polymorphism.

MĔTȦKĪNĔ'SĬS, that part of the metaphases in karyokinesis which includes only the separation of the threads.

MĔT'ȦMĔR, applied by Sachs to a Phyton (which see), or one of any number of similar parts connected in a series.

MĔTȦMÔR'PHŌSĬS, the varied development of members of the same morphological value, resulting from their adaptation to different functions. Compare TRANSFORMATION.

MĔT'ȦPHĀSĒS, a term applied by Strasburger to those portions of karyokinesis which include the longitudinal splitting of the threads, the formation of the mother-star, and the subsequent separation of the segments to form the daughter-stars.

MĔT'ĂPLĂSM, a name given by Hanstein to that part of the protoplasm which holds the formative material. Used by some synonymously with Deutoplasm.

MĔTĂSPĔR'MOŬS, see ANGIOSPERMOUS.

MĔTĂS'TĂSĬS, see METABOLISM.

MĔTĒŎR'ĬC, applied to flowers whose opening or closing is influenced by the weather.

MĔTŒ'ÇIOŬS, see HETERŒCIOUS.

MĔTŒ'ÇĬŞM, see HETERŒCISM.

MĔTŎX'ĒNOŬS, see HETERŒCIOUS.

MĬĂŞMĂT'ĬC, pertaining to infectious organisms floating in the atmosphere. See INFECTION.

MĬÇĔL'LĂ (pl. **Mĭçĕl'læ**), a term proposed by Nägeli for certain structural particles which serve for the building up of all plant and animal tissues. They are considered to be aggregations of molecules which, like crystals, are able to increase or diminish in size without chemical change.

MĪCRĂN'DRE, see DWARF-MALE.

MĪ'CRŌBE, a micro-organism, animal or vegetable. In use restricted chiefly to plants, especially the Schizomycetes, or bacteria and their allies. See MICROPHYTE.

MĪ'CRŌÇŸST, a resting state of certain swarm-cells in Myxomycetes.

MĪCRŌGŌNĬD'ĬŬM (pl. **Mīcrōgōnĭd'ĭă**), applied to gonidia of smaller size than others produced by the same species;—they are usually male gonidia, as the so-called androspores of Œdogonium. In pteridophytes and their allies they are usually called microspores. See MICROSPORE.

MĪCRŎG'RĂPHŸ, the description or study of microscopic objects; micrology.

MĪCRŌPHŸL'LOŬS, having small leaves.

MĪ'CRŌPHŸTE, any plant, such as the bacteria and smaller species of fungi, which can be clearly seen only by means of a microscope.

MĪ'CRŌPŸLE, the nearly closed foramen as it exists in the ripened seed.

MĪCRŌSŌ'MĂ (pl. **Mīcrōsō'mătă**), a name proposed by Strasburger for a class of granules in the protoplasm which have a high degree of refringency and are deeply stained by hæmatoxylin.

MĪCRŌSPŌRĂN'ĠĬŬM (pl. **Mīcrōspōrăn'ġĭă**), a sporangium containing microspores.

MĪ'CRŌSPŌRE, a small male spore, as in Rhizocarpæ, Selaginellæ, and Isoetæ. See MICROGONIDIUM. Compare MACROSPORE.

MĪ'CRŌSTŌME, a small mouth or orifice.

MĪCRŌSTŸ'LOŬS, applied to that form in heterostyled plants in which the flowers have short styles and long filaments. The expressions "long-styled," "medium- or mid-styled," and "short-styled" are preferable to Macrostylous, Mesostylons, and Microstylous, and are more frequently used.

MĪCRŌZŌŌGŌNĬD'ĬŬM, a motile microgonidium, as in Ulothrix, differing from a macrozoögonidium in size and in having the power of conjugation.

MĪ'CRŌZȲME, a microscopic organism capable of producing fermentation.

MĬD'DLE LÁMĔL'LÁ, the portion of a lignified cell-wall which occupies the place of the primary membrane. It usually appears as a fine dividing line between adjoining cells. Compare MIDDLE LAMINA.

MĬD'DLE LĂM'ĬNÁ, the portion of a lignified cell-wall between the middle lamella and the inner lamina.

MĬD'RĬB, the large central vein of a leaf, which usually exists as a ridge continuous with the petiole.

MĬLK, see LATEX.

MĬLK-SĂC, a form of laticiferous vessel existing in some species of maple.

MĬLK-SĂP, see LATEX.

MĬLK-VĔS'SĔL, see LATICIFEROUS VESSEL.

MĬMĔT'ĬC, pertaining to mimicry.

MĬM'ĬCRȲ, special resemblance to another object, serving usually for protection, as when plants bear a marked resemblance in color to the soil in dry regions, thus being less liable to be seen and eaten by animals.

MĬN'ĬĀTE, vermilion-colored.

MĪ'ŌPHȲLLȲ, see MEIOPHYLLY.

MĪŌSTĔM'ŌNOŬS, see MEIOSTEMONOUS.

MĪTŌ'SĬC, pertaining to, characterized by, or exhibiting mitosis; mitotic.

MĬTŌ'SĬS, (1) the splitting of the chromatin of the nucleus (karyomitosis) or the subdivision of any minute granular body found in protoplasm; (2) a figure occurring during mitosis as a result of that process.

MĪTŎT'ĬC, see MITOSIC.

MĪ'TRE-SHĀPED, see MITRIFORM.

MĬT'RĬFÔRM, conical and slightly narrowed toward the mouth like a pope's mitre. Also the same as Cucullate, but not slit upon one side; mitre-shaped.

MĬXED ĬNFLŌRĔS'ÇENÇE, one in which each partial inflorescence develops in a reverse order from that of the general inflorescence, as in Compositæ where the general inflorescence is usually centrifugal, the terminal "flower" opening first, while each head is centripetal, the marginal florets being first developed.

MĬXED VĔS'SĔLS, those having thickenings in their walls of more than one kind, as both spiral and annular thickenings.

MŌ'BĬLE, movable.

MŎL'LĬS, soft.

MŎNÀDĔL'PHOŬS, having the filaments free, and united in a ring around the pistil, as in most Malvaceæ.

MŌNĂN'DROŬS, having but a single stamen.

MŌNĂN'THOŬS, having but one flower upon a plant or peduncle.

MŎNĂS'TĔR, see MOTHER-STAR.

MŌNĒ'ÇIOŬS, see MONŒCIOUS.

MŎN'GRĔL, properly, a cross. Usually applied in the United States to inferior mixed breeds resulting from unknown or heterogeneous crossing.

MŌNĬL'ĬFÔRM, necklace-shaped; cylindrical, and contracted at regular intervals so as to resemble a string of beads. Compare NODOSE.

MONO-, a Greek prefix, meaning one or single.

MŎN'ŌCÄRP, an annual or other plant that fruits but once.

MŌNŌCÄR'PĔLLĀRY̆, of one carpel; monogynous.

MŌNŌCÄR'PĬAN, see MONOCARPIC.

MŎNŌCÄR'PĬC, bearing fruit but once. Compare POLYCARPIC.

MŎNŌCÄR'POŬS, said of a flower in which the gynœcium forms but one ovary, whether simple or compound. Compare POLYCARPOUS.

MŎNŌÇĔL'LŪLAR, see UNICELLULAR.

MŎNŌÇĔPH'ÀLOŬS, in single heads, or having one head. Applied to an ovary with but one style, or to flowers disposed in single umbels or other clusters.

MŌNŌCHĀ'SĬAL, in single dichasia. See DICHASIUM and COMPOUND DICHASIUM.

MŎNŌCHĀ'SĬŬM, a uniparous cyme. Compare DICHASIUM.

MŎNŌCHLÀMY̆D'ĒOŬS, having but one floral envelope or perianth-whorl. As this is usually clearly a calyx, and as in doubtful cases it is considered so, the term is often equivalent to Apetalous. Compare ACHLAMYDEOUS.

MŎNŌCHRŌ'MĬC, see UNICOLOR.

MŌNŎC'LĬNOŬS, see HERMAPHRODITE. Compare DICLINOUS.

MŎNŌCŎTY̆LĔD'ONOŬS, having but one cotyledon or seed-leaf. A term applied to endogens.

MŎNŌÇY̆C'LĬC, of a single whorl.

MŌNŒ'ÇIOŬS, having stamens and pistils in separate flowers on the same plant.

MŌNŒ'ÇIOŬSLY̆ PŌLY̆G'ÀMOŬS, having perfect and separated flowers on the same plant. Compare DIŒCIOUSLY POLYGAMOUS.

MŌNŎG'ÀMOŬS, see HOMOGAMOUS.

MŌNŎG'Y̆NOŬS, having but one pistil or style. There may be several carpels.

MŌNŌĪ'COŬS (obs.), see MONŒCIOUS.

MŌNŌLŎC'ŪLAR, see UNILOCULAR.

MŌNŎM'ĒROŬS, of one part, as a flower having but one organ of each kind, one pistil, one stamen, etc.; applied also to tubers having but one internode.

MŎNŌPĔT'ALOŬS, see GAMOPETALOUS.

MŎNŌPHY̆L'LOŬS, see GAMOPHYLLOUS.

MŎN'ŌPŌDE, see MONOPODIUM.

MŌNŌPŌD'ĬĂL, having a single main axis, as ordinary plants. Compare SYMPODIAL and DICHOTOMOUS.

MŎNŌPŌ'DĬŬM, an axis of growth which continues to elongate at the apex, while lateral structures of like kind are produced beneath it in acropetal succession. Compare SYMPODIUM and DICHOTOMY.

MŌNŎP'TĒROŬS, one-winged.

MŎNŌSĔP'ALOŬS, see GAMOSEPALOUS.

MŎNŌSĪ'PHŎNOŬS, said of a frond in Florideæ when it consists of but a single row of cells.

MŎNŌSPĒR'MOŬS, one-seeded. Compare POLYSPERMOUS.

MŌNŎS'TĬCHOŬS, in one vertical rank.

MŎNŌSTY̆'LOŬS, having but one style.

MŎNŌSY̆MMĔT'RĬCAL, capable of division in but one direction into similar halves, as the flower of the pea; zygomorphons. Compare SYMMETRICAL and POLYSYMMETRICAL.

MŎNŌTHĂL'ĂMOŬS, unilocular; applied to galls and rarely (Tuckerman) to the apothecia of lichens.

MŎNŌTHĂL'MĬC, derived from a single flower, as ordinary fruits. Compare POLYTHALMIC.

MŌNŌTȲP'ĬC, said of a genus which contains but one species, or of the species itself in such a case.

MŌN'STĔR, a plant or animal having any marked abnormal development in form; monstrosity. See SPORT and TERATOLOGY.

MŎNSTRŎS'ĬTȲ, see MONSTER.

MŎN'STROŬS, developed in a remarkably abnormal manner. The term has no reference to size.

MŎP-HĔAD'ĔD, said of a tree without a leader and with an unusually large number of small branches.

MÔRPHŌĠĔN'ĔSĬS, the production of morphological characters.

MŎRPHŎL'ŌĠȲ, the science of the homologies and metamorphoses of members. It treats, for example, of the different forms which leaves assume in different plants, and in the same plant to serve different purposes.

MŎS'CHĀTE, having an odor like that of musk.

MOTHER-CELL, one from which another is derived.

MOTHER-STÄR, a stage of karyokinesis in which the nuclear threads lie in the equator of the nucleus, usually in the form of loops, with their free ends pointing away from the centre; monaster; nuclear disk; nuclear plate; equatorial plate.

MŌT'TLED, covered with partially confluent dots, or with blotches of varying intensity. Applied mainly to fruits.

MOUN'TAĬNOŬS, see ALPESTRINE.

MOVE'MENTS OF VĀRIĀ'TION (Peffer), see ALLASSOTONIC MOVEMENTS.

MŪÇĔD'ĬNOŬS, mould-like.

MŪ'CĬLĀĠE, dissolved vegetable jelly; any slimy vegetable product.

MŪCĬLĂĠ'ĬNOŬS, slimy, or yielding mucilage; mucous.

MŪ'COŬS, see MUCILAGINOUS.

MŪ'CRŌ, a short sharp point.

MŪ'CRŌNĀTE, terminating abruptly in a short stiff point.

MŪCRŎN'ŪLĀTE, diminutive of Mucronate; ending abruptly in a minute stiff point.

MŪLE, see HYBRID. The term Mule is by some restricted to genus-hybrids and by others to sterile hybrids of any origin. The term is now useless in botany.

MŬLTĂN'GŪLAR, having more than four angles; polygonal.

MŬLTĬ-, a Latin prefix denoting many.

MŬLTĬÇĬP'ĬTAL, many-headed; applied to a root or root-stock from which several stems arise.

MŬLTĬCŎS'TĀTE, having many ribs proceeding from the base of the leaf.

MŬLTĬCŬS'PĬDĀTE, having many cusps or points.

MŬLTĬDĔN'TĀTE, having many teeth.

MŬLTĬFĀ'RĬOŬS, (1) diversified, or composed of many diverse parts; (2) arranged in many ranks or rows.

MŬLTĬF'ĔROŬS, producing fruit several times in a season.

MŬL'TĬFĬD, cut about half way to the midrib into many segments; many-cleft.

MŬLTĬFLŌ'ROŬS, many-flowered.

MŬLTĬFŌ'LĬĀTE, having numerous leaflets.

MŬLTĬJŪ'GĀTE, having many pairs of leaflets.

MŬLTĬLĂT'ĔRAL, many-sided.

MŬLTĬLŎC'ŪLAR, many-celled: applied mainly to ovaries or pericarps.

MŬLTĬLŎC'ŪLAR SPŌRE, see COMPOUND SPORE.

MŬLTĬNŪ'CLĒĀTE, having more than one nucleus in a cell.

MŬLTĬPĂR'TĪTE, divided into many parts.

MŬL'TĬPLE, compound.

MŬL'TĬPLE CŌRŌL'LȦ, one having more than one whorl of petals, as in "double" flowers.

MŬL'TĬPLE FRUIT, see COLLECTIVE FRUIT.

MŬL'TĬPLE-PRĪ'MĀRY̆ ROOT, one having several main divisions from the crown, as the fascicled root of dahlia.

MŬL'TĬPLĬCĀTE FLOW'ER, see DOUBLE FLOWER.

MŬLTĬRĀ'DĬĀTE, having many rays.

MŬLTĬRȦMŌSE', having many branches.

MŬLTĬSĔP'TĀTE, having numerous septa.

MŬLTĬSĒ'RĬAL, in several horizontal rows. Compare MULTISTICHOUS.

MŬLTĬS'TĬCHOŬS, in several or many vertical rows.

MŪ'RAL, growing upon walls.

MŪ'RĬCĀTE, covered with short, sharp points. Compare SCABROUS.

MŪRĬC'ŪLĀTE, slightly muricate.

MŪ'RĬFŌRM, arranged like courses of bricks or stones in a wall, as the cells in medullary rays.

MŬSCĂR'ĬFÔRM, having long hairs toward the end like an ancient fly-flap, as the styles of some Compositæ. Compare ASPERGILLIFORM.

MŬS'ÇĬFÔRM, moss-like.

MŬSCŎL'ŌĠY̆, see BRYOLOGY.

MŪ'TĬCOŬS, pointless; blunt. Mainly in distinction from Awned or Mucronate.

MŪ'TŪALĬŞM, see SYMBIOSIS.

MŪ'TŪAL PĂRȦSĪ'TĬŞM, see SYMBIOSIS.

MY̆CĒ'LĬAL STRĂND, see FIBROUS MYCELIUM.

MY̆ÇĒ'LĬŬM, the vegetative portion of a fungus, consisting of one or more hyphæ.

MY̆ÇĒTŌĠĔNĔT'ĬC, produced by fungi.

MY̆ÇĒTŌĠĔNĔT'ĬC MĔTȦMÔR'-PHŌSĬS, deformation due to a parasitic fungus.

MY̆ÇĒTŎ'LŌĠY̆, see MYCOLOGY.

MY̆CŎL'ŌĠY̆, the botany of fungi.

MY̆CŌPRŌ'TĒĬN, a term sometimes applied to the peculiar protoplasm of which the putrefactive bacteria are composed.

MY̆CŌRRHĪ'ZȦ, a term applied by Frank to a symbiotic relation supposed to exist between the roots of many plants and the mycelium of certain fungi in soils containing a large amount of humus.

MY̆CŌ'SĬS, the presence of parasitic fungi in a plant or animal, together with the morbid effects of their presence.

MȲCŌPRŌ'TĒĬN, a term sometimes applied to the peculiar protoplasm of which the putrefactive bacteria are composed.

MȲ'CRŌPȲLE, see MICROPYLE.

MȲXÅMŒ'BÆ, zoöspores in Myxomycetes having an amœboid or creeping motion.

MȲX'ŌSPŌRE, a spore produced in the midst of a gelatinous mass without distinct ascus or basidium.

NĀ'KĔD, destitute of the usual covering, as a cell without a wall, a stem without leaves, a flower without floral envelopes.

NĀ'KĔD-SEEDED, see GYMNOSPERMOUS.

NÅNĂN'DROŬS, having short or dwarf male plants. Applied to certain algæ.

NĀ'NĬSM, a dwarf condition.

NĀ'NŬS, dwarf. Compare HUMILIS.

NĀ'PĬFŌRM, turnip-shaped; i.e., depressed-globose and tapering below, like most varieties of turnip.

NĂS'ÇENT, in the earliest rudimentary condition.

NĂS'ÇENT TĬS'SŪE, see MERISTEM.

NĀ'TANT, floating unattached in or upon water. Compare FLUITANT.

NĂT'ŪRAL GRÅFT'ĬNG, see INARCHING.

NĂT'ŪRALĪZED, introduced from another country, but fully established, growing and reproducing itself spontaneously. Compare ADVENTITIOUS.

NĂT'ŪRAL ŌR'DĔR, see ORDER.

NĂT'ŪRAL SȲS'TĔM, the system of classification introduced by A. L. de Jussieu and now in general use, which takes into account all parts of the plant, and seeks to place each species or other group nearest to those which in all respects it most resembles. Compare ARTIFICIAL SYSTEM.

NAU'TĬFŌRM, see NAVICULAR.

NÅVĬC'ŪLAR, boat-shaped, or like the hull of a ship, as the glumes of many grasses; naviculoid; nautiform; cymbæform; cymbiform; scaphoid.

NÅVĬC'ŪLŌID, see NAVICULAR.

NĔB'ŪLŌSE, clouded.

NĔCK, the line of junction between root and stem (collar); the upper tapering end of a bulb; the narrow part of a pear, toward the base; the upper end of the sheathing petiole in grasses; the upper part of the tube of a gamopetalous corolla (throat); the prolonged apex of a pyrenocarp; the outer portion of an archegonium containing the canal, etc.

NĔCK'LĀÇE-SHĀPED, see MONILIFORM.

NĒCRŎG'ĒNOŬS, hastening or producing death. A term formerly applied to certain parasitic fungi.

NĔC'TAR, a sweet secretion by some part of a flower.

NĔC'TAR-GUĪDE, a term applied to various color-marks on flowers which seem designed to indicate to insects the locality of the nectar; nectar-spot; nectar-mark; honey-guide; honey-spot; pathfinder.

NĔCTARĬF'ĒROŬS, secreting nectar, or having a nectary.

NĔCTĀ'RĬŬM, see NECTARY.

NĔC'TAR-MÄRK, see NECTAR-GUIDE.

NĔCTARŌTHĒ'CA, a spur or other receptacle containing a nectary. (Obs.)

NĔC'TAR-SPŌT, see NECTAR-GUIDE.

NĔC'TARY, the part of a flower which secretes nectar. Applied especially to spur-shaped appendages to the petals containing nectar, as in the columbine, and sometimes used for similar organs which contain no nectar.

NEE'DLE-SHĀPED, see ACEROSE and ACICULAR.

NĔG'ATĬVE ĠĒŌT'RŌPĬSM, see APOGEOTROPISM.

NĔG'ATĬVE HĔLĬŌT'RŌPĬSM, see APHELIOTROPISM.

NĔG'ATĬVELY HĒLĬŌTRŎP'ĬC, see APHELIOTROPIC.

NĔG'ATĬVELY RHĒŌTRŎP'ĬC, said of an organ (usually a root) when it tends to grow in a direction opposite to that of a current of water in which it is placed, as the roots of maize. (Jönsson.)

NĔMATHĒ'ÇIŬM (pl. **Nĕmathē'çia**), a wart-like group or mass of tetraspores in Florideæ.

NĔM'ATŌID, thread-like; filamentous.

NĒ'MĒOŬS, composed of threads or filaments; filamentose. (Obs.)

NĔM'ŌRŌSE, growing in groves.

NEPH'ROID, see RENIFORM.

NĔRV'ĀTE, see NERVED.

NĔRVĀ'TION, the manner in which the nerves of a leaf are arranged; venation.

NĔRVE, a term formerly in general use for the veins of monocotyledons, now mainly used for the veins (when moderately developed) on the floral envelopes of grasses. Compare VEIN and RIB.

NĔRVED, having nerves instead of veins; nervate.

NĔRVŌSE', abounding in nerves.

NEST'LING, see NIDULANT.

NĔT'TĔD, see RETICULATED.

NEŪRĀ'TION, see NERVATION.

NEŪ'TĔR, having neither stamens nor pistils; neutral. Said of a plant or flower.

NEŪ'TRAL, see NEUTER.

NEW SPĒ'ÇIĒŞ. A species is considered new when its name and description are published for the first time. The botanical name is then followed by the words "new species," or their abbreviation ("n. s." or "n. sp." in English, "sp. nov." in Latin), without the name of the author, the writer of the description being understood to be the author of the name. When the name is republished in any manner the species is no longer considered new, and the name of its author is then appended.

NĬD'ŪLANT, embedded in pulp or other loose material, or partly enclosed in a receptacle; nestling.

NĪ'DŬS, a suitable place for a spore or seed to germinate.

NĪ'ĠĔR, black and either glistening or slightly tinged with gray. Compare ATER.

NĪGRĔS'ÇENT, becoming black or blackish; nigricant.

NĬG'RĬCANT, see NIGRESCENT.

NĬT'ĬD, smooth and shining; bright; lustrous. Compare LUCID and LÆVIGATE.

NĬT'ĬDŬS, see NITID.

NĬV'ĒOŬS, snow-white; pure white. A clearer and purer white than Candidus.

NŎCTÛR'NAL, lasting through a night, or occurring at night.

NŌD'DĬNG, see CERNOUS.

NŌDE, the place on a stem where one, two, or more leaves are attached; any knot or swelling; a point of intersection, as of threads in reticulated protoplasm.

NŌDĬF'ĔROŬS, bearing nodes.

NŌDŌSE', knotted; with swollen joints; swollen at intervals. Compare MONILIFORM.

NŎD'ŪLE, a small knot or rounded body.

NŌD'ŬLŌSE, diminutive of Nodose.

NŌR'MAL, according to rule; usual; of the ordinary type or structure.

NŌSŎL'ŌĠY, VEGETABLE, see VEGETABLE NOSOLOGY.

NŌ'TĀTE, marked by colored spots or lines.

NŌTŌRHĪ'ZAL, see INCUMBENT.

NŌ'TŌTRĬBE, said of an irregular flower when arranged so that the pollen from the stamens strikes the back of a visiting insect. Compare STERNOTRIBE and PLEUROTRIBE.

NŪCĂMĔNTĀ'ÇEOŬS, nut-like, or pertaining to nuts.

NŪCĂMĔN'TŬM (obs.), see AMENT.

NŪÇĔL'LŬS, the nucleus of an ovule.

NŪÇĬF'ĔROŬS, nut-bearing.

NŪ'ÇĬFÔRM, nut-shaped; nucamentaceous.

NŪ'CLĒAR, pertaining to a nucleus.

NŪ'CLĒAR BĂR'RĔL, a stage in karyokinesis, immediately preceding the nuclear spindle.

NŪ'CLĒAR DĬSK, see MOTHER-STAR.

NŪ'CLĒAR FĪ'BRĬLȘ, see SPINDLE-FIBRES.

NŪ'CLĒAR FĬL'ĂMENT, a general term for the chromatin of a nucleus, which in karyokinesis appears as segments of a filament, and in the resting state as a filamentous network; chromatic filament. See CHROMOSOME.

NŪ'CLĒAR PLĀTE, see MOTHER-STAR.

NŪ'CLEAR SPĬN'DLE, a spindle-shaped achromatic figure in the nucleus during cell-division, consisting of slender filaments which cross the equator or extend toward it from the poles. It begins to appear in the skein stage, and is completed in the mother-star stage.

NŪ'CLĒAR STÄR, see ASTER.

NŪ'CLĒAR THRĔADȘ, see SPINDLE-FIBRES.

NŪ'CLĔĀTĔD, having a nucleus.

NŪ'CLĒĬN, usually employed in the sense of chromatin. By some authors there is believed to be a distinct chemical substance, a form of protoplasm, which is termed nuclein.

NŪ'CLĒŌLĀTĔD, having a nucleolus.

NŪ'CLĒŌLE, see NUCLEOLUS.

NŪCLĒ'ŌLŬS (pl. Nūclē'ōlī), any distinct body of considerable size within a nucleus—not a definite and constant organic structure. There may be more than one.

NŪ'CLĒŌPLĂȘM, the more fluid protoplasm of the nucleus between the nuclear threads; karyoplasm.

NŪ'CLĒŬS, (1) an organized structure within the living cell by means of which cell-division takes place. It is usually spherical in form, richer in protoplasm, and of higher refractive power than the remaining cell-contents; (2) the part of the ovule containing the embryo-sac (nucellus); (3) any organic centre, as the hilum of a starch-grain.

NŪ'CLĒŬS OF THE EMBRYO-SAC, see SECONDARY NUCLEUS.

NŪCŪLĀ'NĬŬM, a term sometimes applied to fruits like the medlar which contain several nut-like seeds or carpels.

NŪ'CŪLE, (1) the female organ in Chara; (2) any small nut-like seed or pericarp.

NŪDE, see NAKED.

NŪDĬCAU'LOŬS, having no leaves on the stem.

NŪ'MĔROŬS, too many to be readily counted; indefinite.

NÛRSE (Hort.), a shrub or tree which serves for the temporary protection of a younger tree or plant.

NŬT, the fruit of certain trees and shrubs, consisting of a hard shell enclosing the seed;—also applied to other small nut-like fruits, as those of Carex, Rumex, and Lithospermum. These latter are more properly called nutlets. See GLANS.

NŪ'TANT, see CERNOUS.

NŪTĀ'TION, the motion of a flower, leaf, or growing stem in following the apparent movement of the sun from east to west during the day; simple nutation. Compare CIRCUMNUTATION.

NŬT'LĔT, a small nut, or nut-like seed or fruit, as many achenia.

NŪTRĬ'TION includes the absorption, elaboration, distribution, and final assimilation of plant-food resulting in growth.

NŬX, see NUT.

NȲCTŌTRŎP'ĬC, applied to the movements of organs which occur at night, or upon the approach of night, producing what is called the "sleep" of plants.

NȲCTŎT'RŌPĬSM, the so-called "sleep" of plants.

OB-, inversely.

ŌBCLĀ'VĀTE, clavate in shape, but attached by the large end.

ŌBCŌMPRĔSSED', flattened in the reverse of the usual direction, as when a legume is flattened so that the dorsal and ventral sutures approach each other. Compare DEPRESSED.

ŌBCŌN'ĬC, conical, but attached at the apex instead of the base; inversely conical. Also applied improperly in pomology to a fruit which has the form of a very short or flattened cone.

ŌBCŌN'ĬCAL, see OBCONIC.

ŌBCŌR'DĀTE, inversely heart-shaped, i.e., with the attachment at the small end.

ŎBDĬPLŌSTĔM'ŌNOŬS, diplostemonous, with the stamens of the outer whorl opposite to the petals.

ŎBĬM'BRĬCĀTE, (1) having overlapping scales directed downward or backward; (2) having the outer scales in an involucre progressively longer than the interior ones.

ŌBLĂN'ÇĒŌLĀTE, lanceolate in form, but tapering toward the base instead of toward the apex.

ŌBLĀTE', flattened or depressed at the ends or poles. Compare PROLATE.

ŌB'LĬGĀTE, necessary; essential. Compare FACULTATIVE.

ŎB'LĬGĀTE-PĂR'ĂSĪTE, an organism to which a parasitic life is indispensable for the attainment of its development. Compare FACULTATIVE PARASITE.

ŌBLĬQUE', (1) occupying a position between horizontal and erect; (2) having one side extending farther forward than the other; (3) having one side more developed than the other, as many leaves and some fruits; unsymmetrical. In pomology, Oblique is used in the second sense only, the third being called Angular.

ŌBLĬTĔRĀ'TION, see SUPPRESSION.

ŌB'LŌNG, longer than wide, with nearly parallel sides. Compare OVAL.

ŌBŌ'VAL, sometimes used improperly for Obovate.

ŌBŌ'VĀTE, ovate with the attachment at the narrow end.

ŌBŌ'VOĪD, ovoid, and attached by the small end. Sometimes used incorrectly for somewhat obovate.

ŎB'SŌLĔTE, suppressed, or very rudimentary; scarcely apparent. See ABORTIVE.

ŎBTŪSE', having a rounded end or apex; blunt. Compare ACUTE.

ŌBVĂL'LĀTE, appearing as though surrounded by a wall.

ŌBVĔRSE', inverse; reverse;—as a leaf which is narrowest at the base.

ŎB'VŌLŪTE, a modification of Convolute in which the entire halves of adjoining organs successively overlap or enfold each other. Half-equitant is one form of this condition.

ŌÇĔL'LĀTE, see OCELLATED.

ŌÇĔL'LĀTĔD, having a circular patch of color with a spot of another color within it like an eye; ocellate; oculate.

ŌÇĔL'LŬS, a circular spot, resembling an eye.

ŌCHRĀ'ÇEOŬS, brownish yellow; yellowish.

Ō'CHRĔĂ (pl. **Ō'chrēæ**), see OCREA.

Ō'CHRĔĀTE, see OCREATE.

ŎCHRŌLEŪ'COŬS, yellowish white; whiter than Ochraceous.

Ō'CRĔĂ (pl. **Ō'crēæ**), a sheath formed by a stipule or pair of stipules united around the stem, as in sycamore.

Ō'CRĔĀTE, having ocreæ.

ŎCTĂG'ŸNOŬS, see OCTOGYNOUS.

ŎCTĂM'ĔROŬS, having the parts in eights, as a flower with eight petals, eight stamens, etc.

ŎCTĂN'DROŬS, having eight stamens.

OCTO- (or **OCTA-**), (from the Greek,) eight.

ŎCTŎG'ŸNOŬS, having eight pistils or styles; octagynous.

ŎCTŌLŎC'ŪLAR, having eight cells or compartments in a fruit or pericarp.

ŎCTŌPĔT'ALOŬS, having eight petals.

ŎCTŌRĀ'DĬĀTE, having eight rays, as some ligulate flowers.

ŎCTŌSPĔR'MOŬS, eight-seeded.

ŎC'TŌSPŌRE, an eight-fold tetraspore. Formerly applied to the oögonium of *Fucus vesiculosus*, which contains eight oöspheres.

ŎC'TŌSPŌROŬS, containing eight spores, as most asci.

ŎC'ŪLĀTE, see OCELLATED.

ŎDD'LŸ PĬN'NĀTE, see IMPARIPINNATE.

ŎFFĬÇ'ĬNAL, used in medicine or the arts.

ŌFF'SĔT, a short lateral branch or stolon proceeding from the base of the plant, which serves or may serve for propagation. It usually takes root of itself. Lateral bulbs are also called offsets.

ŌFF'SHOOT, an offset, or lateral branch.

OIL-TUBE, see VITTA.

ŌLĒĂĠ'ĬNOŬS, oily.

ŎLĔRĀ'ÇEOŬS, esculent; used for food; applied mainly to garden vegetables.

ŎLĬGĂN'DROŬS, having few stamens; oligostemonous.

ŌL'ĬGÄRCH, applied to a fibrovascular cylinder having few rays, i.e., containing or representing few fibrovascular bundles.

ŎLĬGŌPHȲL'LOŬS, having few, or comparatively few, leaves.

ŎLĬGŌSPĔR'MOŬS, few-seeded.

ŎLĬGŌSTĔM'ŌNOŬS, see OLIGANDROUS.

ŎLĬVĀ'ÇEOŬS, dusky green.

ŌMNĬV'ŌROŬS, applied to a parasitie fungus which attacks many kinds of plants.

ŌM'PHÅLŌDE, see OMPHALODIUM.

ŌMPHÅLŌ'DĬŬM (pl. **Ŏmphålō'diå**), the central part of the hilum, containing the vessels which enter the raphe or chalaza.

ONE-SĬD'ĔD, see SECUND, UNILATERAL, and HOMOMALOUS

ŌNTŎĠ'ĔNȲ, the study of the development of an individual through all its stages. Compare PHYLLOGENY.

Ō'ŌÇȲST, a female organ or oögonium of doubtful nature.

ŌŌG'ÅMȲ, the conjugation of gametes which are dissimilar in form. Compare ISOGAMY.

ŌŌGŌ'NĬŬM (pl. **Ōōgō'niå**), the female sexual organ in Oösporeæ before fertilization, containing one or more oöspheres, as in Peronospora.

ŌŌ-NŪ'CLĒŬS, the nucleus of an oösphere. Compare SPERMONUCLEUS.

Ō'ŌPHŌRE, the first or sexual stage or generation in plants having an alternation of generations, as ferns; oöphyte. Compare SPOROPHORE.

ŌŌPHŌRĬD'ĬŬM (pl. **Ōōphōrĭd'iå**) (obs.), see MACROSPORANGIUM.

Ō'ŌPHȲTE, see OÖPHORE.

Ō'ŌSPĔRM, see OÖSPORE.

Ō'ŌSPHĔRE, the oöspore previous to fertilization; i.e., the cell or protoplasmic mass which after fertilization becomes the oöspore.

Ō'ŌSPŌRĂNĠE, see OÖSPORANGIUM.

ŌŌSPŌRĂN'ĠĬŬM (pl. **Ōōspōrăn'ġiå**), an old term for oögonium.

Ō'ŌSPŌRE, a fertilized oösphere; oösperm. As a result of fertilization the oösphere, thus changed to an oöspore, takes on a firm cell-wall and acquires the power of germination.

ŌPĀQUE', having a dull surface; neither transparent nor shining.

Ō'PEN, applied to fibrovascular bundles which always contain cambium capable of further growth, as in exogens. Compare CLOSED.

ŌPĔR'CŪLAR, like an operculum; operculate.

ŌPĔR'CŪLĀTE, having an operculum. Compare DEOPERCULATE.

ŌPĔR'CŪLŬM, (1) a transversely dehiscent lid or cap, like that covering the capsule of many mosses; (2) the lid of a pitcher-shaped leaf.

ŌPĬS'THŌDAL, posterior; applied to the inner or posterior opening of a stoma. Compare EISODAL.

ŎP'PŌȘITE, opposed in position. Said, for example, of leaves when situated over against each other in pairs at the same height upon the stem. One floral organ, however, is *opposite* another when it is superposed to it, i.e., when it stands between it and the axis. Compare ALTERNATE.

ŎPPŎȘĬTĬFLŌ'ROŬS, having opposite peduncles.

ŎPPŎȘĬTĬFŌ'LĬOŬS, (1) having opposite leaves; adversifolious; (2) situated opposite a leaf, as the tendrils in the grape.

ŌP'PŌȘITELY̆-PĬN'NĀTE, having leaflets in pairs.

ÔRBĬC'ŪLAR, circular in outline, or nearly so; orbiculate.

ÔRBĬC'ŪLĀTE, see ORBICULAR.

ŌR'DĔR, a division of a class containing one or more genera; family.

ÔR'DĬNAL, pertaining to an order.

ŌR'GAN, any part having a special function, as a leaf or stamen. Compare MEMBER.

ÔRGĂN'ĬC ÇĔN'TRE, a point or axis around which growth takes place. It may not remain the mathematical centre.

ÔRGÂNŎĠ'ĔNY̆, the origin and development of organs; organogenesis.

ÔRGÂNŌG'RÂPHY̆, the study of organs.

ÔRGÂNŌPLĂS'TĬC, having the power of producing organs, as *organoplastic* tissue.

ŌR'GANȘ OF RĔPRŌDŬC'TION, those whose chief function is to serve directly or indirectly in the production of seeds or spores. In the higher plants the stamens and pistils are usually indicated by the term.

ŌR'GANȘ OF VĔĠĔTĀ'TION, those which minister to the growth of the whole plant: roots, stem, and leaves.

ŌRĬĔNTĀ'TION, strictly, facing the east. Used in horticulture for the setting of transplanted trees facing the same way as they stood originally. Used in botany for assuming any definite direction with respect to the points of the compass or to the axis of the plant or organ.

ÔR'MŌGŎN, see HORMOGONIUM.

ÔRTHŌPLŌ'ÇEOŬS, having conduplicate cotyledons. (Obs.)

ŌRTHŌS'TĬCHY̆, a vertical row of organs. Compare PARASTICHY.

ÔRTHŎS'TŌMOŬS, having a straight opening.

ŌRTHŌT'RŌPAL, see ORTHOTROPOUS.

ŌRTHŌTRŌP'ĬC, growing erect, as stems in contrast with leaves, etc. Compare PLAGIOTROPIC.

ÔRTHŎT'RŌPOŬS, (1) applied to an ovule or seed in which the hilum coincides with the chalaza and the micropyle occupies the apex opposite. Compare ANATROPOUS. (2) Applied to the embryo it means the opposite of antitropous, i.e., having the radicle next the hilum, as in anatropous seeds.

ŎS'ÇĬLLĀTĬNG, see VERSATILE.

ŎS'MŌSE, the mutual diffusion of liquids of different densities through a membrane. It is the peculiar property of osmose that the denser fluid absorbs from the other through the membrane more than it gives in exchange. The imbibition of fluids by roots is due partially to osmose (endosmose), the fluid within the plant being denser than that in the soil. The outward diffusion of fluid from the plant into the soil, which takes place in a feeble degree, is termed Exosmose.

ŎSMŌ'SĬS, see OSMOSE.

ŎS'SĒOŬS, bony in texture.

ŌS'SĬFĬED, become like bone, as a peach-stone.

ŎS'TĬŌLE, a small orifice, as that in a perithecium for the discharge of spores.

ŎSTĪ'ŌLŬM, see OSTIOLE.

OUT'ĔR GLŪMES, see EMPTY GLUMES.

Ō'VAL, about twice as long as broad, with regular curved outlines; broadly elliptical.

ŌVĀ'RĬŬM (pl. **Ōvā'rĭȧ**), see OVARY.

Ō'VȦRY̆, the part of a pistil containing the ovules or seeds.

Ō'VĀTE, like a longitudinal section of an ordinary hen's egg, with the attachment, if any, at the broad end. Compare OVOID.

Ō'VĬFÔRM, see OVOID.

Ō'VOID, of the shape of a hen's egg, and attached, if at all, at the large end. Compare OVATE.

Ō'VŪLE, the seed previous to fertilization.

ŌVŪLĬF'ĔROŬS, bearing ovules.

Ō'VŬM (pl. **Ō'vȧ**), see OÖSPHERE.

PĂCHY̆PHY̆L'LOŬS, thick-leaved.

PĂD, a popular name for the leaf of a water-lily.

PĂG'ĬNȦ (pl. **Păg'ĭnǣ**), the surface of an organ, especially the upper or lower surface of a leaf.

PĀINT'ĔD, having colors disposed in streaks of unequal intensity. (Rare.)

PÂIRED, see CONJUGATE.

PĀLÆĀ'ÇEOŬS, see PALEACEOUS.

PĂL'ĀTE, the convex base of the lower lip of a personate corolla, as in snap-dragon.

PĀLE, see PALET.

PĀ'LĔȦ (pl. **Pā'lēǣ**), (1) the inner bract of a flower in grasses (palet); (2) one of the bracts or scales upon the receptacle in Compositæ. Also used by some for Ramentum.

PĀLĔĀ'ÇEOŬS, chaff-like; bearing chaff or chaffy scales; consisting of chaff.

PĀLĔŌBŎT'ȦNY̆, see FOSSIL BOTANY.

PȦLĔ'ŌLȦ (pl. **Pȧlē'ōlǣ**), see LODICULE.

PĀLĒŌPHY̅TŎL'ŌĠY̆, see FOSSIL BOTANY.

PĂL'ĔT, the inner bract or chaff of a flower in grasses; formerly known as the upper palet; palea. Compare GLUME.

PĂLĬSĀDE' ÇĔLLS, elongated cells perpendicular to the epidermis in the upper side of most leaves.

PȦLLĔS'ÇENT, somewhat pale.

PĂL'LĬD, pale; wan; deficient in color.

PÄLM, three inches.

PĂL'MĀTE, lobed or divided so that the sinuses all point toward the apex of the petiole, either moderately as in the maple-leaf, or more deeply, when it is sometimes improperly called *pedate,* or so as to make the leaf compound when it is *digitate.*

PĂL'MĀTĔD, see PALMATE.

PĂL'MĀTELY̆ CLĔFT, having the sinuses in a palmate leaf reaching about half way to the base; palmatifid.

PĂL'MĀTELY̆ CŎM'POUND, having the leaflets all attached to the apex of the petiole; digitate.

PĂL'MĀTELY̆ DĬVĪ'DĔD, cleft to the base in a palmate manner, but not compound; palmatisect.

PĂL'MĀTELY̆ LŌBED, having the lobes directed toward the apex of the petiole.

PĂL'MĀTELY̆ NĔRVED, sometimes applied to palmately veined leaves in endogens; palm-nerved; palminerved.

PĂL'MĀTELY̆ PÄRT'ĔD, cleft nearly to the base in a palmate manner.

PĂL'MĀTELY̆ VEINED, having the principal veins proceeding from the apex of the petiole, as in palmate leaves; palm-veined; palmiveined; radiate-veined.

PĂLMĂT'ĬFĬD, see PALMATELY CLEFT.

PĂLMĂT'ĬLŌBED, see PALMATELY LOBED.

PĂLMĂTĬPÄR'TĪTE, see PALMATELY PARTED.

PĂLMĂT'ĬSĔCT, see PALMATELY DIVIDED. Sometimes improperly used for Palmately Parted.

PĂL'MĬNĔRVED, see PALMATELY NERVED.

PÄLM-NĔRVED, see PALMATELY NERVED.

PÄLM-VEINED, see PALMATELY VEINED.

PĂL'ŪDĬNE, see PALUSTRINE.

PĂLŪ'DĬNOŬS, see PALUSTRINE.

PĂL'ŪDŌSE, see PALUSTRINE.

PĂLŬS'TRĬNE, growing in marshes or swamps; paludine; paludinous; paludose; uliginose; uliginous.

PĂNCHE, have pale faint stripes.

PĂN'DŪRĀTE, see PANDURIFORM.

PĂNDŪ'RĬFÔRM, obovate, with one or two large sinuses or concavities in each side, like the body of a violin; fiddle-shaped.

PĂNĠĔN'ĔSĬS, a theory proposed by Charles Darwin to account for the reproduction of every part of an organism. It is, that each separate part of the individual throws off minute reproductive gemmules, which are capable of multiplication, and may be transmitted from generation to generation. Spores and seeds are abundantly supplied with these gemmules from all parts of the organism, and therefore specially fitted for reproduction. If other parts can serve for propagation, it is because gemmules from the remaining parts are present;—if a cutting "strikes root," for example, it is because it contains root gemmules. Atavism is caused by dormant gemmules from a previons generation. A modification of this theory proposed by De Vries is that each cell contains all the essential elements for the reproduction of the plant, the protoplasm containing a vast number of self-reproductive particles (pangens) which transmit the characters of the individual.

PĂN'ĬCLE, a branched raceme, as in oats. Panicles are either determinate or indeterminate in their mode of inflorescence. If the former they are called "cymose;" if the latter, "botryose." They have a variety of shapes, the typical form being loose and spreading, with repeatedly branched peduncles, as in many grasses. Panicles of certain forms have received special names. A compact panicle like that of lilac is called a thyrsus; a spike-shaped panicle like that of wheat is usually called a spike, or compound spike; a corymb might be called a level-topped centripetal panicle, and a cyme a similar centrifugal panicle.

PĂN'ĬCLED, see PANICULATE.

PȦNĬC'ŪLĀTE, resembling a panicle; furnished with panicles; arranged in panicles; panicled.

PȦNĬC'ŪLĀTĔD, see PANICULATE.

PȦNĬC'ŪLĬFÔRM, panicle-shaped.

PĂN'NȦRY̆, pertaining to bread; suitable for making bread.

PĂN'NĬFÔRM, thick and spongy, like coarse cloth.

PĂN'NŌSE, (1) having a felted coat of hairs. Compare FLOCCOSE and MANICATE. (2) Having the texture of coarse cloth; panniform.

PĂNSPĔR'MĬȦ, the theory that seeds and spores are everywhere present, ready to be developed upon favorable opportunity.

PĀ'PĔRY̆, see CHARTACEOUS.

PȦPĬLĬŌNĀ'ÇEOŬS, having flowers, as in the Papilionaceæ, a tribe of the Leguminosæ or pea family; i.e., with a large upper or posterior petal called the vexillum, two lateral petals called wings, and two small inferior petals more or less united into a carina or keel.

PȦPĬL'LȦ (pl. **Pȧpĭl'læ**), a small soft, elongated teat-shaped protuberance. Compare MAMILLA and WART.

PĂP'ĬLLĀRY̆, having or resembling papillæ. Compare VERRUCOSE.

PĂP'ĬLLĀTE, having papillæ; papilliferous. Compare MAMILLATE.

PȦPĬLLĬF'ĔROŬS, see PAPILLATE.

PȦPĬL'LĬFÔRM, in the form of a papilla.

PĂP'ĬLLŌSE, papillate or papillary; especially, covered with papillæ.

PĂPPĬF'ĔROŬS, bearing a pappus, as the seeds of thistle.

PĂP'PĬFÔRM, like a pappus.

PĂPPŌSE', pappiferous or pappiform.

PĂP'PŬS, the calyx in Compositæ, especially when developed in a hairy or feathery manner to effect the dispersion of the seed. Compare COMA.

PĂP'ŪLȦ (pl. **Păp'ūlæ**), see PAPILLA.

PĂP'ŪLOŬS, see PAPILLOSE.

PĂPY̆RĀ'ÇEOŬS, papery; chartaceous.

PĂRȦCŌRŎL'LȦ, see CROWN.

PĂR'ȦÇY̆ST, a term proposed by Tulasne, but not generally adopted, for De Bary's term Pollinodium.

PĂRȦHĒLĬŎT'RŌPĬSM, the power which certain leaves have when placed in bright sunlight of placing their surfaces parallel to the rays of light; diurnal sleep. Compare DIAHELIOTROPISM.

PĂR'ALLĔL NERVED, see PARALLEL VEINED.

PĂR'ALLĔL VEINED, having the veins or the principal ones parallel, or merely diverging: opposed to Reticulate veined and characteristic of the veins (or nerves) of endogens. In some exogens, however, as the beech, the principal lateral veins are parallel, but in exogens the term *parallel veined* is not applied, the condition being expressed in examples like this by the special term *pinnately veined.*

PĂRÀMĪ'TŌM, a name given by Flemming to the more fluid portion of the cell-substance which is contained in the meshes of the *mitom* or network of threads. It is the *paraplasma* of Kupffer.

PĂRÀNĒ'MÀTÀ (sing. **Părànē'mà**), paraphyses among the spores of Fucus and other algæ. (Obs.) See PARAPHYSES.

PĂRÀNŪCLĒ'ŌLŬS (pl **Părànūclē'ōlī**), one of the additional or secondary nucleoli in a nucleus, when there are more than one.

PĂRÀNŪ'CLĒŬS (pl. **Părànū'clēī**), an object resembling an additional nucleus—generally adjoining the true nucleus and in some cases budded off from it; lateral nucleus.

PĂRÀPĔT'ALOŬS, standing by the side of a petal, as stamens in some Rosaceæ.

PĂRÀPHȲL'LŬM (pl. **Păràphyl'là**), a foliaceous expansion which is not a true leaf, as the large stipules of the pea. Especially one of the leaf-like scaly hairs among the true leaves in mosses.

PÀRĂPH'ȲSĒS (sing. **Pàrăph'ȳsis**), sterile filaments among reproductive bodies of various kinds, as those among the archegonia and antheridia in certain mosses, among the asci in Ascomycetes, and among the basidia in Basidiomycetes.

PĂR'ÀSĪTE, a plant which grows upon or within another living body, from which it derives a part or the whole of its nourishment. Compare SAPROPHYTE.

PĂR'ÀSĪTE - SĂP'RŌPHȲTE, a parasite which kills its host, then continues to feed upon it.

PĂRÀSĪT'ĬC CĂSTRĀ'TION, sterility of a reproductive organ caused by a parasite.

PĂR'ÀSĪTĪSED, affected by a parasite; having a parasite growing upon or within it.

PĂR'ÀSĪTĪSM, the condition of being a parasite.

PĂRÀSTĀ'MĔN, see STAMINODIUM.

PĂRÀSTĒ'MŎN, see STAMINODIUM.

PÀRĂS'TĬCHȲ, a whorl, or spiral row of organs. Compare ORTHOSTICHY.

PĂR'ÀSTȲLE, an abortive style.

PĂRÀTŎN'ĬC, depressing or retarding: opposed to stimulating. Applied mainly to certain spontaneous movements of plants, such as the "sleep" of leaves, induced by the withdrawal of the usual stimulus, in distinction from movements caused by mechanical or artificial irritation. Sometimes applied to all movements of organs which are produced by the presence or absence of external stimuli. The term is also applied to the condition of plants when unusually insensitive to stimuli of any kind. Compare PHOTOTONIC.

PARENCHYMA, all tissue composed of cells which do not have tapering extremities. Especially soft cellular tissue like that of pith and the pulp of leaves. Compare PROSENCHYMA and SCLERENCHYMA.

PARENCHYMATOUS, pertaining to, containing, or resembling parenchyma; spongy; porous.

PA'RIES (pl. **Parī'etēs**), the wall of any enclosing body.

PARI'ETAL, belonging to or situated upon the wall, as ovules or seeds upon the wall of the pericarp.

PARIPIN'NATE, see ABRUPTLY PINNATE.

PARŒ'CIOUS, having male and female "flowers" in mosses placed close together. (Lesquereux and James.)

PART'ED, divided nearly to the base.

PARTHENOGEN'ESIS, the occasional production of true spores or seeds without fertilization. Compare APOGAMY.

PAR'THENOSPORE, a spore in certain algæ resembling a zygospore, but produced without conjugation.

PAR'TIAL, secondary or subordinate.

PAR'TIAL IN'VOLUCRE, one which subtends a subordinate part of an inflorescence, as a partial or secondary umbel.

PAR'TIAL PEDUN'CLE, a branch of an inflorescence (pedicel) bearing more than one flower.

PAR'TIAL PET'IOLE, a division of a petiole bearing one or more leaflets; petiolule.

PAR'TIAL UM'BEL, one of the parts or secondary umbels of a compound umbel; umbellule; umbellet.

PART'IBLE, capable of being readily divided, but not dividing spontaneously.

PAR'TITE, see PARTED.

PARTI'TION, see SEPTUM.

PAR'VUS, relatively small.

PASTEUR'ISM, a term applied by some writers to the protective or prophylactic inoculation of the attenuated virus of certain diseases, especially of hydrophobia, as devised by Louis Pasteur. The older term Vaccination, however, is being very properly extended to all operations involving this principle. Compare PASTEURIZATION.

PASTEURIZA'TION, the preservation of wines or other fermented liquids by destroying the fungi and their spores that would produce further and deleterious changes. This is effected by prolonged heating to a moderate degree (about 140° F.) for the purpose of inducing the germination and subsequent destruction of the spores which are present.

PATEL'LA (pl. **Patel'læ**), a form of sessile apothecium in lichens. (Obs.)

PATEL'LIFORM, kneepan-shaped; in the form of a watch-crystal or shallow saucer, especially if thickened like a kneepan.

PAT'ENT, spreading nearly at right angles.

PATH'FINDER, see NECTAR-GUIDE.

PATHOGEN'IC, disease-producing; pathogenetic.

PATHOL'OGY, see VEGETABLE PATHOLOGY.

PAT'ULOUS, slightly spreading.

PAUCI-, a Latin prefix: few.

PAUCIFLŌ'ROŬS, few-flowered.

PEÂR-FŌRM, see PEAR-SHAPED.

PEÂR-SHĀPED, ovoid or obovoid, and contracted at the sides like the Bartlett pear. Compare TEAR-SHAPED.

PĔC'TĬNĀTE, having close narrow divisions like the teeth of a comb. Applied also by De Bary to two series of fibrovascular bundles when the members of each alternate with those of the other, like the teeth of two combs.

PĔD'ĀTE, palmately parted or divided into somewhat linear lobes, with two strong lateral lobes from each of which somewhat linear divisions arise, as in the maidenhair fern and *Viola pedata*. Compare PALMATE.

PĔD'ĀTELY̆ CLĔFT, see PEDATIFID.

PĔD'ĀTELY̆ VEINED, having no mid-vein, but with two strong lateral veins, from each of which others originate which extend toward the apex; pedatinerved. Compare PALMATELY VEINED.

PĔDĂT'ĬFĬD, divided in a pedate manner about half-way to the base; pedately cleft.

PĒDĂT'ĬNĔRVED, see PEDATELY VEINED.

PĔDĂTĬPÄR'TĪTE, pedately veined and lobed with the sinuses reaching nearly to the principal veins.

PĔD'ĬÇĔL, a branch of an inflorescence, supporting one or more flowers. Also applied to a little stalk or support of any kind. In orchids it is a portion of the pollinium in some genera which is derived from the rostellum and connects the caudicles with the retinaculum. Compare FOOTSTALK and PEDUNCLE.

PĔD'ĬÇĔLLĀTE, having a pedicel; pedicelled.

PĔD'ĬÇĔLLED, see PEDICELLATE.

PĒDĬC'ŪLĀTE, see PEDICELLATE.

PĒDŬN'CLE, the stalk of an inflorescence, whether of one flower or more. Compare PEDICEL and PARTIAL PEDUNCLE.

PĒDŬN'CLED, see PEDUNCULATE.

PĒDŬN'CŪLAR, pertaining to or serving as a peduncle; growing upon or attached to the side of a peduncle, as a *peduncular* tendril.

PĒDŬN'CŪLĀTE, having a peduncle, as a *pedunculate* flower or inflorescence; peduncled.

PĔG, a lip or projection at the lower end of the hypocotyl in Cucurbitaceæ which serves in germination to hold the seed-coats down while the cotyledons are withdrawn.

PĒLĀ'ĠIAN, see PELAGIC.

PĒLĂĠ'ĬC, pertaining to the deep sea; widely dispersed through the ocean; pelagian.

PĔL'LĬCLE, a distinct, firm, and in some cases separable, outer layer of the cortex in lichens and some other fleshy fungi.

PĔLLŪ'ÇĬD, translucent.

PĒLŌ'RĬA, a regular condition in flowers which are usually irregular. See REGULAR and IRREGULAR PELORIA.

PĔL'TA (pl. **Pĕl'tæ**), an obsolete term for certain target-shaped apothecia in lichens.

PĔL'TĀTE, shield- or target-shaped; circular, and fixed by the lower surface, especially the centre of it.

PĔL'TĀTELY̆ VEINED, having veins radiating in all directions from the summit of the petiole, as in a peltate leaf.

PĔL'VĬFÔRM, shaped like the cavity of the pelvis; basin-shaped. (Rare.) See ACETABULIFORM and COTYLIFORM.

PĔN'ÇĬLED, marked with fine distinct lines as if with a pencil.

PĔN'ÇĬLLED, see PENCILED.

PĔND'ENT, hanging directly downward. Compare CERNOUS and DROOPING.

PĔN'DŪLOŬS, hanging downward more than cernons, nearly vertically.

PĔNĬÇĬL'LĀTE, bordered or tipped with fine hairs like those of a camel's-hair pencil. Compare PENCILED.

PĔNĬÇĬL'LĬFÔRM, like a camel's-hair pencil.

PĔN'NĀTE, see PINNATE.

PĔN'NĬFŌRM, in the form of a feather or plume.

PĔN'NĬNĔRVED, see PINNATELY VEINED.

PENTA-, a Greek prefix: five.

PĔNTȦCÄR'PĔLLĀRY̆, having an ovary with five carpels.

PĔNTȦCŎC'COŬS, of five cocci.

PĔNTȦÇY̆C'LĬC, in five whorls.

PĔNTĂG'Y̆NOŬS, of five pistils or styles.

PĔNTĂM'ĔROŬS, of five parts: said of a flower having five organs in each whorl. Sometimes written 5-merons.

PĔNTĂN'DROŬS, having five stamens.

PĔNTȦPĔT'ALOŬS, of five petals.

PĔNTȦPHY̆L'LOŬS, five-leaved; applied mainly to the parts of the calyx; pentasepalous.

PĔNTĂP'TĔROŬS, five-winged.

PĔNTȦSĔP'ALOŬS, of five sepals; pentaphyllous.

PĔNTȦSPĔR'MOŬS, five-seeded.

PĔNTĂS'TĬCHOŬS, in five vertical ranks.

PĔ'PŌ, a cucurbitaceous fruit, as a melon or gourd. It is indehiscent, fleshy, inferior, polycarpellary, and has a thick, more or less hardened, rind.

PĔRȦPHY̆L'LŬM, an accrescent foliaceous calyx, as in quince. (Obs.)

PĔRCŬR'RENT, running throughout, as a midrib through the entire length of a leaf to the extreme apex.

PĔRĔN'NĬAL, adj., living more than two years.

PĔRĔN'NĬAL, n., a plant which lives more than two years.

PĔRĔN'NĬAL HĔRB, a plant that lives from year to year, but dies to the ground or nearly so at the close of each season.

PĔR'FĔCT, containing both stamens and pistils; hermaphrodite; bisexual. Compare COMPLETE.

PĔRFŌ'LĬĀTE, said of a leaf which is united around the stem at its base. Compare CONNATE.

PĔR'FŌRĀTE, pierced with small round holes. Compare PERTUSE and PUNCTATE.

PĔRGȦMĒ'NĒOŬS, see PERGAMENTACEOUS.

PĔRGȦMĔNTĀ'ÇEOŬS, like parchment. (Rare.) See CHARTACEOUS.

PĔR'ĬĂNTH, the calyx and corolla taken together, or calyx alone when there is no corolla. Used mainly where calyx and corolla are much alike, as in lilies. Formerly applied to the calyx only, or to calyx and involucre.

PĔRĬĂN'THĬŬM, see PERIANTH.

PĔRĬĂNTHŌMĀ'NĬȦ, an unnatural multiplication of sepals, bracts, etc.

PĔR'ĬBLĔM, the primordial cortex, i.e., the layer of nascent cortex beneath the dermatogen.

PĔRĬCĂM'BĬŬM, see PHLOËM-SHEATH.

PĔR'ĬCÄRP, the wall of a ripened ovary or seed-vessel. Sometimes improperly applied to husks or other protective coverings surrounding the fruit.

PĔR'ĬCHÆTH, the involucre of leaves around the base of the seta (or sessile capsule) in mosses; also around the archegonia in Marchantia, etc.; perichætium.

PĔRĬCHÆ'TIAL, belonging or pertaining to the perichæth.

PĔRĬCHÆ'TĬŬM, see PERICHÆTH.

PĔRĬCLĀ'DĬŬM, a sheathing petiole in exogens, as in many Umbelliferæ. (Rare.)

PĔRĬCLĪ'NAL, said of cell-walls or any lines when parallel with the outer surface of the plant or organ. Compare ANTICLINAL. The annual rings of a trunk, for example, extend in a *periclinal* direction, and the medullary rays in an *anticlinal.*

PĔRĬCLĬN'ĬŬM, a name for the involucre in Compositæ.

PĔR'ĬCȲCLE, the outer portion of the central or fibrovascular cylinder.

PĔR'ĬDĔRM, see EPIPHLŒUM. Also applied to the cell-wall or any other enclosing membrane.

PĔRĬDĪ'ŌLŬM (pl. **Pĕrĭdī'ōlȧ**), diminutive of Peridium: a secondary or interior peridium containing a hymenium, as in Crucibulum.

PĒRĬD'ĬŬM (pl. **Pērĭd'ĭȧ**), the covering of almost any closed spore-case, or fungus-fruit, as of a puff-ball.

PĔR'ĬGŌNE, see PERIANTH.

PĔRĬGŌ'NĬŬM (pl. **Pĕrĭgō'nĭȧ**), see PERIANTH. Sometimes used for Perichæth.

PĔRĬĠȲN'ĬŬM (pl. **Pĕrĭġȳn'ĭȧ**), a term applied to various appendages of unusual or doubtful character surrounding the pistil, as the sac-like envelope (utriculus) surrounding the ovary in Carex. Also used for Perichæth in Hepaticæ and mosses.

PĒRĬĠ'ȲNOŬS, adherent to the calyx and free from the ovary: said of stamens.

PĔRĬPĔT'ALOŬS, around the petals.

PĔRĬPHĔR'ĬC, situated around the outside or periphery.

PĔRĬPHLŌ'ĔM, see PHLOËM-SHEATH.

PĔRĬPHŌRĂN'THĬŬM, see PERICLINIUM.

PĔRĬPHȲL'LŬM (pl. **Pĕrĭphȳl'lȧ**), see LODICULE.

PĔR'ĬPLĂSM, a hyaline peripheral layer of protoplasm; ectoplasm. Especially, in Peronosporeæ, a part of the protoplasm in the oögonium and antheridium which does not take part in fertilization. Compare GONOPLASM.

PĒRĬP'TĒROŬS, surrounded by a wing-like expansion.

PĔR'ĬSPĔRM, the albumen of a seed, especially that formed outside the embryo-sac. Sometimes applied also to the seed-coat and to the pericarp. Compare ENDOSPERM.

PĔRĬSPĔR'MĬC, furnished with perisperm.

PĔRĬSPŌRĂN'ĠĬŬM, see ĬNDUSIUM.

PĔR'ĬSPŌRE, a covering of one or more spores, as the cell or sac in which tetraspores are formed.

PĔR'ĬSTŌME, the fringe of teeth around the mouth of the capsule in mosses.

PĔRĬTHĒ'ÇĬŬM (pl. **Pĕrĭthē'çĭà**), the receptacle containing the asci in Ascomycetes. Commonly applied in a general way to the whole ascocarp.

PĔRĬT'RŌPAL, see PERITROPOUS.

PĔRĬT'RŌPOŬS, having the axis of the seed perpendicular to that of the pericarp. (Rare.)

PĔRĬZŌ'NĬŬM, a thin non-silicious membrane of a young auxospore in Diatomaceæ.

PĔRPĔNDĬC'ŪLAR SŸS'TĔM, see FIBROVASCULAR SYSTEM.

PĔRSĬST'ENT, remaining longer than usual, as parts of the flower which remain (either living or dead) until the fruit is ripe, or leaves which die but remain upon the tree during winter.

PĔR'SŌNĀTE, labiate, with the throat nearly closed by a projection in the lower lip called the palate; masked. Compare RINGENT.

PĔRSPĬRĀ'TION, see TRANSPIRATION.

PĔRTŪS'ĀTE, see PERTUSE.

PĔRTŪSE', having slits or holes; perforate.

PĔRTŪSED', see PERTUSE.

PĔR'ŪLÀ (pl. **Pĕr'ūlæ**), see PERULE.

PĔR'ŪLĀTE, having perules.

PĔR'ṲLE, see BUD-SCALE.

PĔT'AL, a leaf of a corolla; flower-leaf.

PĔT'ALED, having petals; petalous: opposed to Apetalous. Used mainly in compounds, as five-*petaled*.

PĔTALĬF'ĔROŬS, bearing petals.

PĔTĂL'ĬFŌRM, shaped like a petal.

PĔT'ALĬNE, see PETALOID.

PĔT'ALŌDE, an organ resembling a petal; a false petal.

PĒTĂL'ŌDŸ, the conversion of other floral organs into petals.

PĔT'ALŌID, petal-like in texture, color, and form; petaline. Compare CORALLACEOUS.

PĔTALŌMĀ'NĬÀ, any unusual development of the petals. Not usually applied to ordinary doubling, but to some remarkable development in size or form.

PĔT'ALOŬS, see PETALED.

PĔT'ĬŌLAR, pertaining to or having a petiole; borne on a petiole, as a *petiolar* gland or tendril.

PĔT'ĬŌLĀRŸ, pertaining to a petiole

PĔT'ĬŌLĀTE, having a petiole; petioled.

PĔT'ĬŌLE, the stem or foot-stalk of a leaf. Compare STIPE and PEDUNCLE.

PĔT'ĬŌLED, see PETIOLATE.

PĔTĬŎL'ŪLAR, pertaining to the petiole of a leaflet.

PĔT'ĬŌLŪLE, see PARTIAL PETIOLE.

PĔZ'ĬZŌID, in the shape of Peziza; cyathiform, or nearly so.

PHÆNŌCĂR'POŬS, having the fruit free from the perianth. (Rare.)

PHÆNŎG'ĀMOŬS, see PHENOGAMOUS.

PHÆNŌLŎĠ'ĬCĂL, see PHENOLOGICAL.

PHÆ'ŌÇȲST, an old term for Nucleus.

PHĀ'LĂNX (pl. Phâlăn'ġēs), a bundle of more or less coherent stamens with broad filaments, as in some Malvaceæ. Compare ADELPHIA.

PHĂN'ĔRŌGĂM, see PHENOGAM.

PHĂNĔRŌGĀ'MĬAN, see PHENOGAMOUS.

PHĂNĔRŌGĂM'ĬC, see PHENOGAMOUS.

PHĂNĔRŎG'ĂMOŬS, see PHENOGAMOUS

PHĔL'LĔM, cork.

PHĔL'LŌDĔRM, green cells beneath the cork formed from the inner layers of the phellogen; cork-cortex.

PHĔL'LŌĠĔN, the inner active growing layers of cork-tissue; cork-cambium; cork-meristem.

PHĒ'NŌGĂM, a flowering plant, i.e., one which produces true seeds; phanerogam. Compare CRYPTOGAM.

PHĔNŌGĀ'MĬAN, see PHENOGAMOUS.

PHĔNŌGĂM'ĬC, see PHENOGAMOUS.

PHĒNŎG'ĂMOŬS, producing true flowers and seeds. Compare CRYPTOGAMOUS.

PHĒNŌLŎĠ'ĬCAL, applied to the blossoming of plants, and other periodical phenomena of plants and animals; phænological.

PHLŌ'ĔM, that portion of a fibrovascular bundle which contains the bast and sieve tissue; leptome. In exogens it is always sharply defined from the remaining portion (xylem) by a layer of cambium. The inner bark is derived from the phloëm, and the wood from the xylem.

PHLŌ'ĔM-RĀY, a ray or plate of phloëm between two medullary rays. It is an outward continuation of a xylem-ray.

PHLŌ'ĔM-SHĔATH, a layer of thin-walled cells surrounding the fibrovascular cylinder next within the cortex—usually better defined in roots than in stems; bast-sheath; periphloëm; pericambium; vascular bundle-sheath.

PHŌRĂN'THĬŬM, see ANTHODIUM.

PHŌTŌ-ĔP'ĬNĂSTȲ, downward curvature due to the presence of light. Compare PHOTO-HYPONASTY.

PHŌTŌ-HȲ'PŌNĂSTȲ, upward curvature due to increased illumination. Photo-epinasty and photo-hyponasty should be distinguished from heliotropism, as the curvatures are in the directions stated, from whatever direction the light comes.

PHŌTŌTĂX'ĬS, taking a definite position with reference to the incident rays of light, as certain desmids and the leaves of the compass-plant, *Silphium laciniatum*.

PHŌTŌTŎN'ĬC, a term applied to the stimulating influence of light upon plants, inducing or increasing irritability and excitability. Thus, when growth which has been arrested by prolonged darkness is restored upon admission of light, the effect is termed the *phototonic* influence of the light; with most organs, however, growth is more rapid in darkness. This effect of light in retarding growth is called its *para-*

tonic influence. See PARATONIC.

PHRĂG'MÀ (pl. **Phrăg'màtà**), a horizontal false dissepiment in a pericarp, as in some species of Cassia. Formerly any false dissepiment, or any dissepiment.

PHȲCŌCȲ'ÀNĪNE, a bluish pigment, found in certain marine algæ.

PHȲCŌĒRȲTH'RĪNE, the reddish pigment in Florideæ.

PHȲCŌL'ŌĠȲ, see ALGOLOGY.

PHȲCŌ'MÀ, the whole plant in algæ, including thallus and reproductive organs. (Obs.)

PHȲCŌMĀ'TẼR, gelatine containing spores in algæ. (Obs.)

PHȲCŌPHÆ'ĪNE, the brown pigment of Fucaceæ and some other algæ.

PHȲCŌXĂN'THĪNE, a buff-colored pigment in diatoms and certain other algæ; diatomine.

PHȲ'LÀ, pl., see PHYLUM.

PHȲLLĀ'RĪĒS, an old term for the bracts forming the luvolucre of the flower-head in Compositæ.

PHȲL'LŌCLĀDE, see PHYLLOCLADIUM.

PHȲLLŌCLĀ'DĬŬM (pl. **Phȳllōclā'dĭà**), a flattened branch which somewhat resembles a leaf, as in Ruscus and Psilotum; cladode; cladodium; cladophyll; phylloclade. Compare PHYLLODIUM.

PHȲLLŌCȲ'ÀNĪN, a bluish pigment which with phylloxanthine forms the green coloring matter of chlorophyll; cyanophyll.

PHȲL'LŌDE, see PHYLLODIUM.

PHȲLLŌ'DĬŬM (pl. **Phȳllō'dĭà**), a dilated petiole taking the place of a blade.

PHȲL'LŌDȲ, the reversion of bracts or floral organs to leaves; frondescence; phyllomorphy.

PHȲL'LŌĠĔN, see PHYLLOPHORE.

PHȲLLŌĠĔNĔT'ĬC, leaf-producing.

PHȲLLŎĠ'ĒNOŬS, growing upon leaves.

PHȲL'LOID, leaf-like.

PHȲLLŌMĀ'NĬÀ, an abnormally abundant growth of leaves, or their production in unusual places. Compare PLEIOPHYLLY.

PHȲL'LŌME, a general term for all organs which are morphologically leaves, as bracts, scales, petals, etc.

PHȲLLŌM'ĬC, pertaining to a leaf or phyllome.

PHȲL'LŌMŌRPHȲ, see PHYLLODY.

PHȲL'LŌPHŌRE, any leaf-bearing organ, especially the leaf-bearing portion of the stem in palms.

PHȲLLŌPŌ'DĬŬM, the branched or unbranched axis of a leaf, as the stipe and rachis of a frond, an ordinary petiole, or any arrangement which serves as an axis or support for the expanded portion or portions of a leaf.

PHȲLLŌTĂX'ĬS, the order of arrangement of leaves upon stems; phyllotaxy.

PHȲL'LŌTĂXȲ, see PHYLLOTAXIS.

PHȲLLŌXĂN'THĬN, a yellow pigment associated with phyllocyanin in the production of chlorophyll; xanthophyll.

PHȲLŎĠ'ĔNȲ, the comparative study of the development of animals or plants. It seeks to trace the origin of species, varieties, etc., and their various organs from preëxistiug forms. It deals with the history of a species or other group, in distinction from Ontogeny, which deals with the history of an individual. See EVOLUTION.

PHY̆'LŬM (pl. **Phy̅'lȧ**), a scale, group, or system of organisms arranged in the manner in which its individuals or subordinate groups have succeeded each other in point of time.

PHY̆ȘĬŌLŎĠ'ĬCAL BŎT'ĂNY̆, the science which treats of how plants live, grow, and perform their various functions; vegetable physiology.

PHY̆TŌĠĔN'ĔSĬS, plant reproduction, germination, and development, or the science which treats of these processes; phytogeny.

PHY̆TŎĠ'ĔNY̆, see PHYTOGENESIS.

PHY̆TŌĠĒŎG'RȦPHY̆, see GEOGRAPHICAL BOTANY.

PHY̆TŎG'RȦPHY̆, descriptive botany, including both the describing and naming of plants. See VEGETABLE TAXONOMY.

PHY̆'TOID, plant-like.

PHY̆'TŌLĪTE, see PHYTOLITH.

PHY̆'TŌLĬTH, a fossil petrified plant; phytolite.

PHY̆TŌLĬTHŎL'ŌGY̆, see FOSSIL BOTANY.

PHY̆TŎL'ŌĠY̆, see BOTANY.

PHY̆TŎL'Y̆SĬS, the change in position undergone by the chlorophyll of a cell in consequence of the alternation of day and night or of the intensity of the sunlight. See APOSTROPHE, EPISTROPHE, and SYSTROPHE.

PHY̆'TŌMĔR, see PHYTON.

PHY̆'TŎN, an internode with a node at its upper extremity which bears one or more leaves, in the axil of each of which may appear one or more buds; phytomer. The name was given by Gaudichaud, who regarded plants as compound individuals made up of successive *phytons*. See METAMER.

PHY̆TŎN'ŌMY̆, see PHYSIOLOGICAL BOTANY.

PHY̆TŎN'Y̆MY̆, see PHYSIOLOGICAL BOTANY.

PHY̆TŌPȦTHŎL'ŌĠY̆, see VEGETABLE PATHOLOGY.

PHY̆TŎT'ŌMY̆, (1) the science of vegetable anatomy; (2) the art or act of plant-dissection. Compare STRUCTURAL BOTANY

PHY̆TŌZŌ'ŎN (pl. **Phytōzō'ȧ**) (obs.), see ANTHEROZOID.

PĪ'LĔATE, like a pileus; having a cap or pileus; pileiform.

PĬLĔ'ĬFŌRM, shaped like the pileus or cap of a mushroom.

PĬLĔ'ŌLŬS (pl. **Pīlē'ōlī**), a little pileus, especially where there are several from the same stem.

PĬLĔŌRHĪ'ZȦ, see ROOT-CAP.

PĪ'LOŬS, see PILOSE.

PĪ'LĔŬS (pl. **Pī'lēī**), the cap of mushrooms. Extended to other similar objects, as the stalked stroma of Claviceps.

PĪLĬF'ĔROŬS, bearing hairs.

PĬL'ĬFÔRM, hair-like.

PĪLĬĠ'ĔROŬS, producing hairs.

PĬLŌSE', (1) covered with long, soft, nearly erect and somewhat distant hairs; (2) having the nature of hair.

PĪ'LŬS (pl. **Pī'lī**), a hair.

PĬN-EȲED, a florist's term for certain flowers which have the style more conspicuous than the stamens. Compare THRUM-EYED.

PĬN'NȦ (pl. Pĭn'næ), a leaflet, or branch of a pinnately-compound leaf.

PĬN'NĀTE, having leaflets borne along a common petiole; pinnately compound. Compare BIPINNATE and TRIPINNATE. See LEAFLET.

PĬN'NĀTĔD, see PINNATE.

PĬN'NĀTELȲ CŌM'POUND, see PINNATE.

PĬN'NĀTELȲ CLĔFT, see PINNATIFID.

PĬN'NĀTELȲ DĔCŌMPOUND', bipinnate, or further divided in a pinnate manner.

PĬN'NĀTELȲ DĬVĪ'DĔD, see PINNATISECT.

PĬN'NĀTELȲ LŌBED, having several lobes of about the same size on each side of an elongated leaf; pinnatilobate.

PĬN'NĀTELȲ PÄRT'ĔD, see PINNATIPARTITE.

PĬN'NĀTELȲ TĔR'NĀTE, see PINNATELY TRIFOLIATE.

PĬN'NĀTELȲ TRĬFŌ'LĬĀTE, trifoliolate, with at least the terminal leaflet distinctly stalked; pinnately ternate.

PĬN'NĀTELȲ VEINED, having one primary vein or midrib from which secondary veins run parallel toward the margin, as in the beech; feather-veined.

PĬNNĂT'ĬFĬD, pinnately veined with marginal divisions reaching about half-way to the midrib.

PĬNNĂTĬLŌ'BĀTE, see PINNATELY LOBED.

PĬNNĂTĬPÄR'TĪTE, having marginal divisions in a pinnate leaf reaching nearly to the base; pinnately parted.

PĬNNĂT'ĬSĔCT, having the lobes of a pinnate leaf divided to the midrib but not petioled.

PĬN'NĬFŌRM, like a feather.

PĬN'NĬNĔRVED, see PINNATELY VEINED.

PĬN'NŪLȦ (pl. Pĭn'nūlæ), see PINNULE.

PĬN'NŪLĀTE, having pinnules.

PĬN'NŪLE, a secondary or other subordinate pinna, as in pinnately compound or pinnately decompound leaves.

PĬP, originally any seed, now sometimes applied to the seeds of the apple and to some other small seeds or seed-like bodies, including the little bulbs of lily-of-the-valley.

PĪ'SĬFŌRM, pea-shaped.

PĬS'TĬL, the central seed-bearing organ of a flower, consisting of one or more united carpels. It consists of the seed-containing portion called ovary, the pollen-receiving part called stigma, and generally an intervening stem called the style. Usually there is but one pistil in a flower, but when, as in the strawberry, there are several distinct bodies as here described seated upon the receptacle, each is properly called a pistil.

PĬS'TĬLLĀRȲ CŌRD, an old and inappropriate name for the conductive tissue of the style. See CONDUCTIVE TISSUE.

PĬS'TĬLLĀTE, said of a flower containing pistils but no fertile stamens.

PĬSTĬLLĬD'ĬŬM (pl. Pĭstĭllĭd'ĭȧ), see ARCHEGONIUM.

PĬSTĬLLĬF'ĔROŬS, bearing pistils, or pistils without stamens.

PĬS'TĬLLŌDY̆, the transformation of other organs into pistils.

PĬT, (1) a small depression, or a thin spot in a cell-wall; (2) the endocarp of a drupe containing the kernel or seed; stone. See BORDERED PIT and BORDERED PORE.

PĬT-CHĀM'BĔR, the cavity of a bordered pit upon one side of the closing membrane.

PĬTCH'ĔR, see ASCIDIUM.

PĬTCH'ĔR-SHĀPED, when applied to a corolla means tubular with a contracted throat, as in Vaccinium; urceolate. Applied to other organs it means shaped more or less like an ordinary pitcher with a lip or spout at the top, as the leaf in the various "pitcher-plants."

PĬTH, the column or cord of soft cellular tissue at the centre of an exogenous stem; medulla.

PĬT'TĔD, marked with small depressions.

PĬT'TĔD VĔS'SĔLS, vessels having thickenings in the form of a network with polygonal meshes, leaving thin spots or pits; dotted ducts. Compare BORDERED PIT.

PĬTŪ'ĬTOŬS, pertaining to pitch or mucus.

PLĂÇĔN'TȦ, the part of the ovary to which the ovules are attached; the tissue from which the sporangia arise in ferns. Compare HYMENIUM.

PLĂÇĔNTĀ'TION, the arrangement of the seeds in the pericarp; the arrangement of the placenta itself.

PLĂÇĔN'TĬFÔRM, in the form of a double concave lens; i.e., like a circular thickened disk depressed in the centre upon both sides.

PLĀĠĬŌTRŎP'ĬC, growing at an angle from the vertical or from the axis either upward or downward. Compare ORTHOTROPIC.

PLĀIN, said of a margin which is not undulate in any degree, though it may be sinuate. (E.S. Goff.)

PLĀIT'ĔD, see PLICATE.

PLĀNE, flat.

PLĀNE OF ĬNSĔR'TION, a plane which passes through the point of insertion of a lateral member, as a leaf, and coincides with the main axis and the axis of the lateral member.

PLĀNE OF SY̆M'MĒTRY̆, any plane which divides an object into symmetrical halves.

PLĂN'ŌGĂMĒTE, a ciliated or otherwise motile coalescing (sexual) cell; zoögamete.

PLANT-CĀNE, a sugar-cane produced directly from seed, in distinction from Ratoon, which see.

PLĂNT'LĔT, a small or young plant.

PLĂNT PȦTHŎL'ŌĠY̆, see VEGETABLE PATHOLOGY.

PLĂŞM (or **Plăş'mȧ**), see PROTOPLASM. Also used for nutritive cell-fluids of all kinds.

PLĂŞ'MȦSŌME, a protoplasmic corpuscle.

PLĂŞMĂT'ĬC, serving for growth; plastic. (Rare.)

PLĂŞMŌ'DĬŬM (pl. **Plăşmō dĭȧ**), a mass of naked multinucleated protoplasm exhibiting amœboid movement; the vegetative body in Myxomycetes.

PLĂŞMŎL'Y̆SĬS, the contraction of protoplasm under the influence of reagents.

PLĂŞMŌLY̆T'ĬC, pertaining to plasmolysis.

PLĂS'TĬC, serving the purpose of growth; plasmic; formative.

PLĂS'TĬD, one of a class of clearly defined protoplasmic granules in the protoplasm of active cells which forms the basis of the chlorophyll and other color-granules, and is also the centre at which starch-grains are produced. For the synonymy of the colorless plastids, see LEUCOPLAST; and for that of the color-plastids, see CHROMATOPHORE.

PLĂS'TĬN, see ACHROMATIN.

PLĂS'TŌID, see RHABDOID.

PLĀTE, see NUCLEAR PLATE and SIEVE-PLATE.

PLȦTEAU', the very short stem which bears the scales in a bulb. Formerly called Corm or Cormus. See CORM.

PLĂTȲCÄR'POŬS, broad-fruited.

PLĂTȲLŌ'BĀTE, broad-lobed.

PLĂTȲPHȲL'LOŬS, broad-leaved.

PLEIO-, a Greek prefix meaning full of, or many.

PLEĪŌMÔR'PHĬṢM, (1) a change of form due to excessive growth of an organ; (2) sometimes used in fungi for Polymorphism.

PLEĪŎPH'ȲLLOŬS, having leaves without apparent buds or branches in their axils. (Rare.) See PLEIOPHYLLY.

PLEĪ'ŌPHȲLLȲ, a state in which there is an abnormal number of leaves from the same point, or an unusually large number of leaflets in a compound leaf. Compare POLYPHYLLY.

PLEĪŌSPĔR'MOŬS, containing an abnormally large number of seeds. Formerly the same as Polyspermous.

PLEĪ'ŌTĂXȲ, a state in which there are an abnormally large number of whorls.

PLĒ'NŬS, full, applied to "double" flowers.

PLĒŌMÔR'PHĬṢM, see PLEIOMORPHISM.

PLĒ'ŎN, a term proposed by Nägeli for those aggregates of molecules which cannot be increased or diminished without changing their chemical nature.

PLĒ'ŌNĂṢM, having any part abnormally numerous. (Rare.)

PLĔ'RŌME, nascent fibrovascular tissue.

PLĔ'RŌME-SHĒATH, the phloëm-sheath in its nascent state.

PLĒSĬŌMÔR'PHOŬS, nearly of the same form.

PLEŪRĔṈ'ЄHȲMȦ, fibrous woody tissue. (Rare.)

PLEŪRŌCÄR'POŬS, see CLADOCARPOUS.

PLEŪRŌDĬS'COŬS, growing upon the sides of the disk, as the ray-flowers in Compositæ.

PLEŪRŌGȲ'RĀTE, having the annulus in ferns placed laterally, as in the genus Trichomanes.

PLEŪRŌRHĪ'ZAL, see ACCUMBENT.

PLEŪ'RŌTRĪBE, said of zygomorphic flowers which have the stamens so placed that an insect entering will receive the pollen upon its side, as in the pea. Compare NOTOTRIBE and STERNOTRIBE.

PLĔX'ŪS, any network.

PLĪ'CȦ, see POLYCLADY.

PLĪ'CĀTE, folded like a fan.

PLĪCĂT'ŪLĀTE, diminutive of Plicate.

PLŪMŌSE', like a feather, as the slender branches of the pappus in thistles, which have a row of fine hairs on each side.

PLŪ'MŪLȦ, see PLUMULE.

PLŪ'MŪLE, the rudimentary stem and leaves between the cotyledons.

PLURI-, a Latin prefix: more than one.

PLŪRĬFŌ'LĬOŬS, several-flowered.

PLŪRĬFŌ'LĬĀTE, having several leaves.

PLŪRĬFŌ'LĬŌLĀTE, having several or many leaflets.

PLŪRĬLŎC'ŪLAR, having more than one cell in an ovary; multilocular. Compare UNILOCULAR.

PLŪRĬPĔT'ALOŬS, see POLYPETALOUS.

PLŪRĬSĔP'TĀTE, having more than one septum.

PNEŪ'MȦTŌDEȘ, upward growths from the roots of palms and some other plants which assist in aëration. The "knees" of the bald cypress were formerly supposed to be of this character.

PŎC'ŪLIFÔRM, deep cup-shaped, with hemispherical base and nearly upright sides. Compare ALVEOLATE and CAMPANULATE.

PŎD, any dry dehiscent fruit; capsule. The term pod is the more popular, and is usually restricted in its use, among botanists as well as others, to capsules of considerable size, especially when somewhat rounded or inflated, as in the milkweed (Asclepias).

PŌDĒ'TĬĬFÔRM, resembling a podetium.

PŌDĒ'TĬŬM (pl. **Pōdē'tĭȧ**), a stalk of an apothecium in lichens. Also applied to the fruit-stalk in Marchantia.

PŌ'DĬŬM, a support for some other part. The stem, for example, is a *podium* for the branches. Used mainly in composition. See MONOPODIUM and SYMPODIUM.

PŎDŌCĂR'POŬS, having a gynophore.

PŎDŌÇĔPH'ȦLOŬS, said of a head of flowers when supported on a distinct peduncle or pedicel.

PŎDŌĠȲN'ĬŬM, see GYNOPHORE.

PŌDŎP'TĒROŬS, having winged peduncles.

PŎD'ŌSPĒRM, see FUNICULUS.

PO̅INT'AL, an old term for Pistil.

PO̅INT'LĔSS, see MUTICOUS.

PO̅INT'LĔTĔD, see APICULATE.

PŌ'LAR BŎD'Ȳ, see POLAR CELL.

POLAR CELL, a portion of a gamete budded off prior to fertilization; apoblast; directive corpuscle; polar body; polar globule. Rare in plants.

PŌ'LAR CÔR'PŬSÇLE, a central mass in each aster of a dividing nucleus.

PŌ'LAR GLŎB'ŪLE, see POLAR CELL.

PŌ'LAR NŪ'CLEŬS, a fourth nucleus in each end of the embryo-sac previous to fertilization. The two polar nuclei unite to form the nucleus of the embryo-sac or "secondary nucleus."

PŌLĂR'ĬTȲ, the state of having distinct poles; the tendency to assume a direction parallel to the poles of the earth, as the leaves of the compass-plant, *Silphium laciniatum*.

PŎL'LĔN, the fertilizing powder, usually yellow, produced in the anthers of flowers. It consists of unicellular grains of definite form, varying according to species, which begin the process of fertilization when placed upon the stigma by an act of germination.

PŎL'LĔNĀTE, see POLLINATE.

PŌLLĔNĀ'TION, see POLLINATION.

PŌL'LĔN-ÇĔLL, sometimes applied to the cells or chambers of the anther which contain the pollen; pollen-sac.

PŌL'LĔN-CHĀM'BĔR, in gymnosperms, a cavity at the apex of the ovule in which the pollen-grains lie during fertilization; pollinic chamber.

PŌL'LĔN-GRĀIN, the usual term for an individual spore, cell, or particle of pollen.

PŎL'LĔNĪZE, see POLLINATE.

PŎLLĔNĬZĀ'TION, see POLLINATION.

PŎL'LĔNŌĪD, used by Bennett & Murray for Antherozoid.

PŌL'LĔN-MȦSS, see POLLINIUM.

PŌL'LĔN-SĂC, the cavity of an anther containing the pollen; pollen-cell.

PŌL'LĔN-SPŌRE, see POLLEN-GRAIN.

PŌL'LĔN-TĔTRȦHĔ'DRŌN (pl. **Tĕtrȧhĕ'drȧ**), see POLLEN-TETRAD.

PŌL'LĔN-TĔT'RĂD, a pollen-mass consisting of four pollen-grains united, either permanently or before fully developed; fourfold pollen-grain; pollen-tetrahedron.

PŌL'LĔN-TŪBE, a thin slender tube which issues from the pollen-grain on its contact with the stigma, which it penetrates until it reaches the ovule where fertilization takes place.

PŌL'LĔX, an inch. (Obs.)

PŎLLĬNĀ'RĬŬM (pl. **Pŏllĭnā'rĭȧ**), see ANTHERIDIUM.

PŎL'LĬNĀTE, to place pollen upon the stigma; pollenate; pollenize.

PŎLLĬNĀ'TIŎN, the placing of pollen upon the stigma—the first stage of fertilization; pollenization.

PŌLLĬN'ĬȦ, pl., see POLLINIUM.

PŎLLĬN'ĬC CHĀM'BĔR, see POLLEN-CHAMBER.

PŎLLĬNĬF'ĔROŬS, pollen-bearing.

PŎLLĬN'ĬŬM (pl. **Pŏllĭn'ĭȧ**), a coherent mass of pollen-grains in certain plants, as orchids and milk-weeds, so arranged as to be conveyed by insects; pollen-mass. For the terminology of the parts supporting a pollinium, see RETINACULUM, CORPUSCULUM, CAUDICLE, and PEDICEL.

PŎLLĬNĬZĀ'TIŎN, see POLLINATION.

PŎLLĬNŌ'DĬŬM, the antherium or male sexual organ in Pyrenomycetes which, either directly or by means of an outgrowth, conjugates with the female organ in fertilization.

POLY-, a prefix derived from the Greek, meaning many.

PŎLȲȦDĔL'PHĬAN, see POLYADELPHOUS.

PŎLȲȦDĔL'PHOŬS, having the stamens united by their filaments into three or more sets. See ADELPHOUS, MONADELPHOUS, and DIADELPHOUS.

PŎLȲĂD'ĔNOŬS, bearing many glands.

PŎLȲĂN'DRĬAN, see POLYANDROUS.

PŎLȲĂN'DROŬS, having twenty or more hypogynous stamens. Compare ICOSANDROUS.

PŎLȲĂN′THOŬS, having many flowers, especially if in one head. A polyanthous involucre, for example, is one investing many flowers.

PŎLȲCÄR′PĔLLĀRȲ, said of a pistil consisting of more than one leaf or carpel.

PŎLȲCÄR′PĬC, fruiting successively; sychnocarpous. Compare MONOCARPIC. Sometimes improperly used for Apocarpous.

PŎLȲCÄR′POŬS, used both for Polycarpic and Apocarpous. Compare MONOCARPIC and MONOCARPOUS.

PŎLȲÇĔPH′ȦLOŬS, bearing many heads.

PŎLȲCLĀ′DĬȦ, see POLYCLADY.

PŎLȲCLĀ′DOŬS, having abnormally numerous branches.

PŎL′ȲCLĀDȲ, an excessive development of twigs or branches; plica. Due either to disease or teratology.

PŎL′ȲCLŌNȲ, an old term for Polyclady.

PŎLȲCŎC′COŬS, of several cocci.

PŎLȲCŎTȲLĔD′ȮNOŬS, having more than two seed-leaves.

PŎLȲCŎTȲLĔD′ŎNȲ, an abnormal increase in the number of cotyledons.

PŎLȲDĔL′PHOŬS, see POLYADELPHOUS.

PŎLȲĔM′BRȲŌNĀTE, having more than one embryo in a seed.

PŎLȲĔM′BRȲŌNȲ, the production, either abnormally or regularly, of more than one embryo in a seed. The term has been restricted to cases where the additional embryos arise without fertilization outside the embryo-sac, but there seems to be no good reason for the restriction.

PŎLȲFLŌ′ROŬS, see MULTIFLOROUS.

PŎLȲGĀ′MĬAN, see POLYGAMOUS.

PŎLȲGȦMŌ-DĬŒ′ÇIOŬS, see DIŒCIOUSLY POLYGAMOUS.

PŎLȲG′ȦMOŬS, producing male and hermaphrodite, or female and hermaphrodite, or male, female, and hermaphrodite flowers on the same or on different individuals; i.e., having both perfect flowers and those of one sex.

PŎLȲG′ŌNOŬS, having many angles, knots, or nodes.

PŎLȲǴȲNŒ̄′ÇIAL, containing the gynœcia of several flowers, as a collective fruit.

PŎLȲǴ′ȲNOŬS, having many styles or pistils.

PŎLȲHĒ′DRŎN (pl. **Pŏlyhē′drȧ**), in Hydrodictyon, a special angular cell with horn-like processes, formed by the swarm-cells produced in the zygospore, and within which a new cœnobium is developed.

PŌLȲM′ĔROŬS, having many parts, or more than one: said of a flower with more than one organ in each whorl, or of a whorl containing more than one organ. Compare MONOMEROUS.

PŎLȲMÔR′PHĬC, see POLYMORPHOUS.

PŎLȲMÔR′PHĬȘM, (1) a condition in which different individuals of the same species have different forms, as in many diœcious plants; (2) the state of passing different stages of existence under distinct forms which might be mistaken for different species, as is the case with heterœcious and some other fungi; pleiomorphism. See METAGENESIS, ALTERATION of Generations, and Heterœcism.

PŎLȲMÔR'PHOŬS, existing under different forms. See above. Also having numerous more or less definite sub-types under a given type.

PŎLȲPĔT'ALOŬS, having distinct petals (opposed to Gamopetalous); apopetalous; dialypetalous; eleutheropetalous; choripetalous.

PŎL'ȲPHŌRE, a common receptacle for many distinct carpels, as that of the strawberry.

PŌLȲPH'ȲLLOŬS, see Polysepalous.

PŎL'ȲPHȲLLȲ, an increase in the usual number of floral organs in a whorl. Compare Pleiophylly.

PŎLȲRHĪ'ZAL, many-rooted.

PŎLȲSĔP'ALOŬS, of two or more distinct sepals; aposepalous; apophyllous. Compare Polypetalous.

PŎLȲSĪ'PHŌNOŬS, said of the thallus in Florideæ when it consists of several parallel rows or filaments of cells.

PŎLȲSPÊR'MOŬS, containing many seeds. Compare Pleiospermous.

PŎL'ȲSPÊRMȲ, the fertilization of a female cell by more than one male ceii.

PŎL'ȲSPŌRE, see Compound Spore.

PŎLȲSPŌ'ROŬS, containing many spores.

PŌLȲS'TĂCHOŬS, having many spikes.

PŎLȲSTĔM'ŌNOŬS, having many more stamens than petals. Compare Isostemonous and Polyandrous.

PŎLȲSȲMMĔT'RĬCAL, capable of division into two symmetrical or equal halves in more than one direction; actinomorphous.

PŎLȲTHĂL'MĬC, derived from more than one flower, as a collective fruit. Compare Monothalmic.

PŌLȲT'ŌMOŬS, having the blade of a leaf distinctly divided into many subordinate parts, but not compound; having the stem forked or divided into many coördinate parts.

PŌLȲT'RĬCHOŬS, bearing numerous hairs.

PŎLȲZȲGŌ'SĬS, the conjugation of more than two cells (gametes).

PŌMĀ'ÇEOŬS, having the appearance or nature of an apple.

PŌME, an indehiscent fruit of more than one carpel, with the seeds enclosed in horny or parchment-like endocarps, and an adnate fleshy calyx, as in the apple.

PŌMĬF'ĔROŬS, pome-bearing.

PŎM'ĬFÔRM, apple-shaped.

PŌMŎL'ŎĠȲ, the department of horticulture which relates to fruits. See Fruit. Compare Carpology.

PŌRE, a small circular opening.

PŌRE-CĂNĂL', the passage through a bordered or other pit between adjoining cells.

PŌRE-CĂP'SŪLE, one from which the seeds or spores escape by a pore or pores.

PŌRĔN'CHȲMÀ (obs.), see Pitted Tissue.

PŌ'RŌSE, pierced with many small circular openings.

PŎRRĔCT', directed outward or forward; outstretched. Compare Arrect.

PŎR'RĔT, see Scallion.

PŎS'ĬTĬVE ĠEŎT'RŌPĬSM, growing toward the centre of the earth—usually called simply

Geotropism. Compare NEGATIVE GEOTROPISM.

PŎS'ĬTĬVE HĒLĬŎT'RŌPĬSM, the same as Heliotropism. Compare NEGATIVE HELIOTROPISM.

PŎS'ĬTĬVELY̆ RHĒŌTRŎP'ĬC, having the direction of growth in a rheotropic organ coincide with that of the stream in which it is placed, i.e., point down-stream. (Jönsson.) Compare NEGATIVELY RHEOTROPIC.

PŎSTĒ'RĬOR, the side of a flower, etc., adjoining the axis or main stem. Compare ANTERIOR and DORSAL.

PŎS'TĬCOŬS, see EXTRORSE.

POUCH, see SILICLE.

POUCH-SHĀPED, like a short bag, as the pod of shepherd's-purse.

PRÆ'CŌX, see PRECOCIOUS.

PRÆFLŌRĀ'TION, see ÆSTIVATION.

PRÆFŌLĬĀ'TION, see VERNATION.

PRÆMŌRSE', see PREMORSE.

PRĀTĔN'SĬS, growing in meadows.

PRĒCŌ'ÇIOŬS, appearing or ripening before the proper or usual time: said of flowers which expand before the leaves, and of plants which flower or fruit much younger than usual, etc.

PRĒFLŌRĀ'TION, see ÆSTIVATION.

PRĒFŌLĬĀ'TION, see VERNATION.

PRĒMÔRSE', abrupt, and irregularly notched at the end as if bitten off. Compare TRUNCATE.

PRĬCK'LE, a small, sharp, stiff outgrowth from the epidermis. Compare SPINE and THORN.

PRĪ'MĀRY̆, chief or first formed. For examples see PRIMARY AXIS, PRIMARY CORTEX, etc.

PRĪ'MĀRY̆ ĂX'ĬS, the main stem.

PRĪ'MĀRY̆ CÔR'TĔX, the true cortex or fundamental tissue of the bark. Compare SECONDARY CORTEX. See PERIBLEM.

PRĪ'MĀRY̆ LĒAVES, see PRIMORDIAL LEAVES.

PRĪ'MĀRY̆ MĔR'ĬSTĔM, the growing tissue of a young organ. Compare SECONDARY MERISTEM.

PRĪ'MĀRY̆ PĔT'ĬŌLE, the midrib of a compound leaf.

PRĪ'MĀRY̆ ROŌT, the central or main root, being a direct continuation of the stem; tap-root.

PRĪ'MĀRY̆ STRŬC'TŪRE, the early structure of a plant or organ after all its distinctive tissues are formed and before any further growth or modification takes place.

PRĪ'MĀRY̆ TĬS'SŪE, (1) tissue in the condition when first formed; (2) that which is formed during the first stage or season of growth.

PRĪ'MĀRY̆ WŎOD, that contained in the fibrovascular bundles of exogens when first developed, before the formation of the cambium ring. Compare SECONDARY WOOD.

PRĪ'MĪNE, the outer coat of the ovule, called testa in the seed. Mirbel applied the term primine to the *inner* coat of the ovule, because first formed, and some other German writers have used it in the same sense.

PRĬM'ĬTĬVE, original: applied, for example, to the original species from which cultivated plants have been derived.

PRĬMŌR'DĬAL, original, or first formed.

PRĬMŌR'DĬAL ÇĔLL, a cell without a cell-wall; naked cell.

PRĬMŌR'DĬAL ĔPĬDĔR'MĬS, the epidermis as it exists when first formed.

PRĪMÔR'DĬAL LĒAVEŞ, the first leaves to succeed the cotyledons. Applied especially to lower leaves which differ considerably from those on the upper portion of the stem. Compare PROTOPHYLL.

PRĬMŌR'DĬAL Ū'TRĬCLE, the outer layer of protoplasm adjoining the cell-wall. The term has with some about the same significance as Ectoplasm, though it usually refers more partienlarly to the immediate surface of the protoplasm (considered as a membrane, though not really one) rather than to a definite outer layer. The term was first used by Mohl and applied to the layer of protoplasm adjoining the cell-wall in cells which are nearly filled with sap. Upon the application of certain reagents the protoplasm contracts from the wall as a sac, the "primordial utricle."

PRĪMÔR'DĬŬM (pl. **Prīmôr'dĭà**), any member or organ in its earliest condition.

PRĬŞMĂT'ĬC, in the form of a prism—with flat, longitudinal faces separated by angles. Applied to stems. Compare TERETE.

PRŌCĂM'BĬŬM, the first formed fibrovascular tissue of an organ before it becomes differentiated into xylem and phloëm. Compare CAMBIUM.

PRŌ'CÄRP, see PROCARPIUM.

PRŌ'CÄRPE (Bornet & Thuret), see PROCARPIUM.

PRŌCÄR'PĬŬM (pl. **Prōcär'pĭà**), in Florideæ, the female organ (archicarp) before fertilization. It consists of a carpogonium, together with the trichogyne and any other accessory part. Compare CYSTOCARP.

PRŎÇ'ĔSS, any projection from a surface.

PRŌCŬM'BENT, see PROSTRATE.

PRŌDŬÇED', prolonged; extended; projected.

PRŌĔM'BRȲŌ, (1) the Suspensor, which see; (2) formerly applied to a prothallus, or to the first result of the germination of any spore; now restricted to special cases, as the rudimentary first stage of the sporophore arising from the oöspore in Characeæ. Compare FROMYCELIUM, PROTONEMA, and PROTHALLUS.

PRŌĔMBRȲŌN'ĬC BRANCHES, short branches sometimes found on the nodes of *Chara fragilis* which resemble the proëmbryos in structure and serve for reproduction.

PRŎG'ĂMOŬS, preceding fertilization: applied to the cell of the pollen-grain which forms the pollen-tube, in distinction from vegetative cells which are also sometimes found.

PRŌGRĔS'SĬVE MĔTĂMÔR'PHŌSĬS, the appearance in place of organs of the usual character of those belonging to a higher or succeeding set, as when petals are replaced by or "converted into" stamens; ascending metamorphosis. Compare RETROGRESSIVE METAMORPHOSIS.

PRŌ'LĀTE, elongated in a polar direction. Compare OBLATE.

PRŌLE, a useless term applied both to Form and Race.

PRŌLĔP'SĬS, accelerated, anticipated, or hurried development, as in the disease known as "peach-yellows," where axillary buds develop into branches the first year.

PRŌLĬFĔRĀ'TION, development in a proliferous manner.

PRŌLĬF'ĔROŬS, developing buds, branches, flowers, etc., from unusual places. Applied, for example, to a flower from which another flower or a branch proceeds, to a leaf from which other leaves or branches arise, to a bulbous plant which abnormally produces bulbs upon the stem above ground, or to any plant which forms young plants in unusual number about its base.

PRŌLĬF'ĬC, fruitful. Sometimes used in the sense of Proliferous.

PRŌLĬFĬCĀ'TION, development in a prolific or proliferous manner; proliferation.

PRŌLĬG'ĔROŬS, bearing reproductive bodies of any kind.

PRŎM'ĬNENT, standing out more than usual, or beyond adjoining parts.

PRŌMȲCĒ'LĬŬM, in Uredineæ and Ustilagineæ, a short and short-lived mycelial growth proceeding from a resting-spore and upon which sporidia are borne.

PRŌ'NĀTE, inclined to grow prostrate; somewhat prone or prostrate.

PRŌNE, lying flat, especially with the face downward; ventricumbent. See PROSTRATE and SUPINE.

PRŌNŪ'CLĒŬS, the nucleus of a gamete. Compare GERM-NUCLEUS.

PRŌP, see FULCRUM.

PRŌPĂC'ŪLŬM, see PROPAGULUM.

PRŌPĂG'ŪLÀ, pl., see PROPAGULUM.

PRŌPĂG'ŪLE, see PROPAGULUM.

PRŌPĂG'ŪLŬM (pl. **Prōpăg'ūlà**), a term applied to runners, offsets, and stolons—especially to a slender runner or stolon terminating in a new plant.

PRŌPĔND'ENT, hanging forward and downward.

PRŎP'ĔR, true or individual. A *proper* calyx, for example, would be the true calyx of an individual flower as opposed to the general calyx (involucre) of a head.

PRŎP'ĔR JŪIÇE, a term formerly used for any characteristic fluid of a plant (especially if thickened) in distinction from the ordinary sap, as the "milk" of milk-weeds. Applied also to the cambium or so-called "descending sap" when in a growing mucilaginous condition.

PRŌPH'ĀSĒŞ, a term proposed by Strasburger for all the phenomena of karyokinesis up to the longitudinal splitting of the threads. Compare METAPHASES and ANAPHASES.

PRŌPHLŌ'ĔM, the first-formed elements of phloëm in a fibrovascular bundle; protophloëm. Applied also to the cylinder of elongated cells with thickened walls containing granular protoplasm found in the seta of certain mosses surrounding the proxylem.

PRŎSCŎL'LÀ, an old term for the retinaculum in orchids.

PRŎSĔN'CHȲMÀ, tissue composed of elongated cells with tapering extremities in the wood and liber. Compare PARENCHYMA.

PROS'TRATE, lying flat upon the ground, but not rooting; procumbent. Compare CREEPING.

PRŌTĂN'DRÓŬS, having stamens which ripen their pollen before the pistils of the same flower are ready for fertilization. Compare PROTOGYNOUS.

PRŌ'TĔĬD, see ALBUMINOID.

PRŌ'TĔĬN, see ALBUMINOID. Formerly considered a distinct substance.

PRŌ'TĔĬN CRȲS'TAL, see CRYSTALLOID.

PRŌ'TĔĬN GRĀIN, see ALEURONE.

PRŌ'TĔN, see PROTENCHYMA.

PRŌTĔN'CHȲMȦ, a term used by Nägeli for all tissue not of the fibrovascular system; proten. Compare EPENCHYMA. These terms are little used.

PRŌTĔRĂN'DROŬS, see PROTANDROUS.

PRŌTĔRĂN'THOŬS, having flowers which appear before the leaves, as in the red maple. Sometimes improperly used in the opposite sense. Compare HYSTERANTHOUS and SYNANTHOUS.

PRŌTĔRŎĠ'ȲNOŬS, see PROTOGYNOUS.

PRŌTHĂL'LĬȦ, pl., see PROTHALLIUM.

PRŌTHĂL'LĬŬM (pl. **Prōthăl'lĭȧ**), see PROTHALLUS.

PRŌTHĂL'LŬS, the thalloid structure resulting from the germination of the spore in ferns and other pteridophytes, upon which the antheridia and archegonia are borne; prothallium. Extended also to the sexual generation of other plants which have an alternation of generations.

PRŌTŌGĔN'ĔSĬS, a term proposed by Rocison for reproduction by budding after the manner of protophytes.

PRŌTŌĠĔN'ĬC, formed in the beginning: said of intercellular spaces which are formed at the time the tissues are beginning to differentiate. Compare HYSTEROGENIC.

PRŌTŎĠ'ȲNOŬS, having pistils which are ready for fertilization before the pollen of the same flower is ripe. Compare PROTANDROUS.

PRŌ'TŌ-MĔR'ĬSTĔM, see PRIMARY MERISTEM.

PRŌTŌNĒ'MȦ (pl. **Prōtōnē'mȧtȧ**), the proëmbryo, or confervoid (often branched) filament first formed from the spore in mosses, and upon which the conspicuous moss-plant is developed by budding.

PRŌTŌPHLŌ'ĔM, see PROPHLOËM.

PRŌ'TŌPHȲL, a cotyledon or other first-formed leaf of a plant; protophyllum. Used mainly in vascular cryptogams. Compare PRIMORDIAL LEAVES.

PRŌTŌPHȲL'LŬM (pl. **Prōtōphȳl'lȧ**), see PROTOPHYL.

PRŌTŌPHȲTŎL'ŌĠȲ, (1) the study of protophytes; (2) fossil botany.

PRŌ'TŌPLĂSM, the nitrogenous fluid of variable composition found in living cells. It is the vital substance into which all food is assimilated and from which all parts of the plant are formed.

PRŌ'TŌPLĂST (Hanstein), the smallest body of protoplasm capable of individual action, either with or without a cell-wall, and either associated in a tissue or independent. It is generally at least a nucleus, together with the protoplasm associated with it. Essentially a cell.

PRŌ'TŌSPŌRE, any spore which develops a promycelium. Applied by Cook to æcediospores. An unnecessary term. Compare SPORIDIUM.

PRŌTŌTHĂL'LŬS, see PROTHALLUS.

PRŌTŌXȲ'LĔM, see PROXYLEM.

PRŌTŌZŎPH'ĬLOŬS, applied to certain water-plants which have the pollen conveyed by minute animals called protozoa.

PRŌTRU̱D'ĬNG, see EXSERTED.

PRŌX'ĬMAL, pertaining to the base, or extremity of attachment. Compare DISTAL.

PRŌXȲ'LĔM, the first-formed xylem in a bundle; protoxylem. The term has also been applied to the layer of water-conducting cells destitute of protoplasm around the central strand in the seta of certain mosses. Compare PROPHLOËM.

PRU̱'ĬNĀTE, see PRUINOSE.

PRU̱'ĬNŌSE, covered with a powdery bloom, like the fruit of most plums; pruinate. Compare GLAUCOUS, HOARY, and CANESCENT.

PRU̱'ĬNOŬS, see PRUINOSE.

PRU̱'NĬFŌRM, plum-shaped.

PRU̱'RĬENT, stinging, as the hairs of nettle.

PSEŪDĂX'ĬS, see SYMPODIUM.

PSEUDO-, a prefix derived from the Greek signifying false or spurious.

PSEŬ'DŌ-BŬLB, see CORM. Especially the corm or fleshy base of the stem in many epiphytic orchids.

PSEŬ'DŌ-BŬLBĬL, an outgrowth which sometimes replaces the ordinary sporangia in ferns and bears antheridia and archegonia.

PSEŬ'DŌCÄRP, any fruit which is not derived exclusively from a single ovary without accessory parts. The fruits of the apple, rose, mulberry, and juniper are pseudocarps.

PSEŪDŌCŌS'TĀTE, said of a leaf in which the veins unite to form an outer vein parallel to the margin, as in Eucalyptus.

PSEŪDŌ-ĠĔ'NŬS, see FORM-GENUS.

PSEŪDŌĠȲ'RĀTE, sometimes applied to the annulus in ferns when it crowns the sporangium, as in the Schizæaceæ. Compare PLEUROGYRATE.

PSEŪDŌMŎNŌCŌTȲLĔD'ŌNOŬS, having two or more consolidated cotyledons. (Obs.)

PSEŪDŌ-PĂR'ĀSĪTE, see EPIPHYTE and SAPROPHYTE.

PSEŪDŌ-PĀRĔṈ'ꞒHȲMĀ, a term applied by De Bary to tissue in fungi which is formed by interlacing and united hyphæ.

PSEŪDŌPĔRĬTHĒ'ÇĬŬM, a false perithecium.

PSEŪDŌ-PĬN'NĀTE, having leaflets (or rather segments) which are not articulated or petioled at their base; pinnatisect.

PSEŪDŌPŌ'DĬŬM (pl. **Pseūdōpō'dĭā**), (1) a stipe or stem of unusual origin, as the leafless upper portion of the stem which supports the capsule in Sphagnum (instead of the true seta which remains undeveloped; (2) one of the temporary changeable branches of a plasmodium, or one of the retractile appendages of the zoöspores (myxamœbæ) of Myxomycetes.

PSEŬDŌRĂM'ŬLŬS, a false or spurious branch found in certain Nostochaceæ, consisting of a younger filament agglutinated for a portion of its length to an older one.

PSEŬ'DŌSPĔRM, an Achene or Caryopsis.

PSEŬDŌSPĔR'MĬC, having a single seed so closely invested by the pericarp that the whole appears like a seed, as in grasses and Compositæ; pseudospermous.

PSEŪDŌSPĒR'MOŬS, see PSEUDOSPERMIC.

PSEŪDŌSȲN'CÄRP, see COLLECTIVE FRUIT. Compare SYNCARP.

PTĒRĬD'ĬŬM, see SAMARA.

PTĔRŌCÄR'POŬS, wing-fruited.

PTĔRŌCAU'LOŬS, having a winged stem.

PTĔ'ROĪD, wing-like.

PTĒRŎP'ŌDOŬS, having a winged petiole.

PTĔRŌSPĒR'MOŬS, having the seeds winged.

PTĔRȲGŌ'NOŬS, having winged expansions on the angles of the stem.

PTŌ'MAĬNE, any alkaloid formed by the activity of bacteria.

PŪ'BĔRTȲ, the period when a plant first begins to blossom.

PŪBĔR'ŬLENT, minutely pubescent; covered with short, soft, rather distant hairs. Compare HOLOSERICEUS.

PŪBĔS'ÇENÇE, soft and rather short hairs; also extended to hairs of all kinds.

PŪBĔS'ÇENT, covered with fine, soft, rather short hairs.

PŪĠĬŎN'ĬFÔRM, dagger-shaped. (Obs.)

PṲL'LEȲ-SHĀPED, see TROCHLEAR.

PŬL'LŬLĀTĬNG, budding; sprouting;—now used only for the budding or sprouting of a cell, a special form of cell-multiplication as seen in yeast, in which a new cell gradually swells out from an older one.

PŬLP, the soft, more or less juicy portion of a fruit, or other plant - substance of similar structure.

PŬLVĒRĀ'ÇEOŬS, dusty or powdery on the surface. Compare PULVERULENT.

PŬLVĔR'ŬLENT, (1) powdery or crumbly; (2) pulveraceous.

PŬL'VĬNĀTE, shaped like a cushion or pillow; having a pulvinus.

PŬL'VĬNĀTĔD, having a pulvinus.

PŬLVĬN'ŬLŬS (pl. **Pŭlvĭn'ūlī**), diminutive of Pulvinus. Applied to various excrescences, and also formerly to the soredia of lichens.

PŬLVĪ'NŬS (pl. **Pŭlvī'nī**), a term applied to various cushion-shaped or wart-like protuberances and swellings, as (1) the projection left by the fall of a leaf in many plants; (2) the enlargement at the base of the petiole in certain Leguminosæ which has the power of contraction, producing the "sleep" of leaves; (3) an enlargement formed by the thickening of the bark at the base of certain twigs, as in arbor-vitæ; (4) a thickened, usually median, portion of the prothallus in ferns bearing the archegonia and antheridia.

PŬ'MĬLŬS, see Humilis.

PŬNC'TĀTE, dotted as if by punctures. The dots may be colored or colorless, superficial or internal, in the latter case sometimes caused by minute oil-glands. Compare Perforate.

PŬNC'TĬFŌRM, in the form of either a dot or point.

PŬNC'TŬM VĔĠĔTĀTĬŌ'NĬS, the extremity of a stem, or other central point where the cells are in the process of division and growth; growing point; apical cone.

PŬN'ĠENT, terminating in a hard sharp point. Compare Mucronate.

PŬR'PŌSE, see Adaptation.

PŬS'TŪLĀTE, bearing pustules or low, blister-like elevations; pustular; pustulose.

PŬS'TŪLE, a slight elevation like a pimple or little blister. Compare Papilla.

PŪTĀ'MĔN, the stone of a drupe, or shell of a nut.

PȲC'NĪDE, see Pycnidium.

PȲCNĬD'ĬŬM (pl. **Pȳcnĭd'ĭà**), a receptacle in Ascomycetes containing stylospores; clinosporangium.

PȲCNŌÇĔPH'ÀLOŬS, having the flowers densely crowded in the inflorescence.

PȲCNŌCŌNĬD'ĬŬM, see Stylospore.

PȲCNŌGŌNĬD'ĬŬM, see Stylospore.

PȲC'NŌSPŌRE, see Stylospore.

PȲCNŎS'TÀCHOŬS, in compact spikes.

PȲŌGĔNĔT'ĬC, pus-forming, as certain bacteria.

PȲRĂM'ĬDAL, either angular and tapering upward, or conical.

PȲRĒ'NÀ (pl. **Pyrē'næ**), a nutlet, or the stone of a small drupe.

PȲ'RĒNE, see Pyrena.

PȲRĒ'NŌCÄRP, (1) the perithecium in Pyrenomycetes; (2) a drupe.

PȲRĒ'NOIDS, minute rounded colorless granules, one or more of which are embedded in the chromatophores of many algæ; amylum bodies; chlorophyll vesicles.

PȲR'ĬFŌRM, see Pear-shaped.

PȲX'ĬDĀTE, resembling a pyxidium or bearing pyxidia.

PȲXĬD'ĬŬM (pl. **Pyxĭd'ĭà**), a capsule which dehisces by a circular transverse line; pyxis. Now restricted to seed-capsules, but formerly applied also to certain spore-capsules, as those of mosses.

PȲX'ĬS (pl. **Pyx'ēs**), see Pyxidium.

QUAD-, or **QUADRI-**, a prefix from the Latin meaning four.

QUADRĂN'GŪLAR, four-angled.

QUADRĬCRU'RAL, having four stems or supports.

QUADRĬDĔN'TĀTE, four-toothed.

QUADRĬDĬĠ'ĬTĀTE, digitate in four divisions.

QUADRĬFĀ'RĬOŬS, in four vertical ranks.

QUAD'RĬFĬD, cleft into four segments half-way to the base or midrib.

QUAD'RĬFOIL, see Quadrifoliate.

QUADRĬFŌ'LĬĀTE, strictly, four-leaved, but used for having four leaflets arising from the apex of the petiole; quadrifoliolate; quadriphyllous; quadrifoil.

QUADRĬFŌ'LĬŌLĀTE, the same as Quadrifoliate, and more precise; i.e., having four leaflets arising from the apex of the petiole.

QUADRĬFÛR'CĀTE, dividing into four branches.

QUADRĬGĔM'ĬNĀTE, growing in fours

QUADRĬJ'ŪGĀTE, having four pairs of leaflets; quadrijugous.

QUADRĬJ'ŬGOŬS, see QUADRIJUGATE.

QUADRĬLŌ'BĀTE, having four lobes, as a leaf.

QUADRĬLŎC'ŪLAR, having four cells in an ovary.

QUAD'RĬNĀTE, see QUATERNATE.

QUADRĬP'ÅRTĬTE, divided to the base or midrib in four parts; four-parted.

QUADRĬPH'ȲLLOŬS, see QUADRIFOLIATE.

QUADRĬVĂL'VŪLAR, of four valves—said of pericarps.

QUĀQUÅVĔR'SAL, directed every way.

QUAR'TĔRĬNG, applied by florists to petals which have an external angle or vacancy between them.

QUAR'TĬNE, a fourth integument in an ovule (if present) counting from the outside. It is really a layer or fold of the secundine or of the nucleus.

QUÅTĒR'NÅRȲ, of four parts. Compare TETRAMEROUS.

QUÅTĒR'NĀTE, growing in fours.

QUĬLLED (Hort.), applied to double flowers in Compositæ when the corollas of the florets, instead of being ligulate, are more or lèss tubular in form like a cornet of paper, as is often seen in the dahlia.

QUIN-, a prefix from the Latin meaning five.

QUĪ'NÅRȲ, in fives or multiples of five.

QUĪ'NĀTE, growing together in fives, as five leaflets from the apex of a petiole.

QUĬNCŬN'ÇIĂL, applied in æstivation to a whorl of five parts, two of which are external, two internal, and one half external and half internal (the typical method of imbricative æstivation when the parts are five). Formerly used also in the sense of five-ranked. In general use the term is applied to objects arranged in squares with one at the centre. In horticulture this arrangement is sometimes termed "false quincuncial," the true quincuncial arrangement being the disposition of objects so that the intervening spaces are all hexagons.

QUĬN'CŬNX, a set of five objects arranged in a quincuncial manner.

QUĬNQUĂN'GŪLAR, five-angled.

QUĬNQUĔCŌS'TĀTE, five-ribbed.

QUĬNQUĔDĔN'TĀTE, five-toothed.

QUĬNQUĔFĀ'RĬOŬS, five-ranked.

QUĬN'QUĔFĬD, in five segments reaching about half-way to the base or margin; five-cleft.

QUĬNQUĔFŌ'LĬĀTE, having five leaflets; strictly, five-leaved.

QUĬNQUĔFŌ'LĬŌLĀTE, the same as Quinquefoliate; i.e., having five leaflets.

QUĬNQUĔJŪ'GĀTE, in five pairs.

QUĬNQUĔLŌ'BĀTE, having five lobes.

QUĬNQUĔLŎC′ŬLAR, having five cells in a pericarp.

QUĬN′QUĔNĔRVED, having two strong veins arising from the midrib on each side near the base; quintuple nerved or veined. Compare QUINQUECOSTATE and QUINQUEVEINED.

QUĬNQUĔP′ÀRTĬTE, divided into five parts nearly to the base.

QUĬNQUĔVĂL′VŬLAR, having five valves in a pericarp.

QUĬN′QUĔVEINED, having five strong veins proceeding from the base of a leaf. The same as Quinquenerved, and usually preferable.

QUĬN′TŪPLE, in fives, or multiples of five.

QUĬN′TŬPLE-NĔRVED, see QUINQUENERVED.

QUĬN′TŬPLE-RĬBBED, see QUINQUECOSTATE.

RĀÇE, a variety, artificial or natural, which reproduces itself from seed.

RÀÇĔME′, an indeterminate inflorescence consisting of single-flowered pedicels arranged along a common axis. Compare PANICLE and SPIKE

RĂÇĔMĬF′ĔROŬS, bearing racemes.

RĂÇ′ĔMŌSE, disposed in racemes or resembling a raceme.

RĂÇ′ĔMŬLE, a small raceme.

RÀÇĔM′ŪLŌSE, producing or disposed in small racemes, or resembling a small raceme.

RÀCHĬL′LÀ, the axis of a spikelet in grasses.

RĀ′CHĬS, a common peduncle or petiole or elongated receptacle, especially (1) the midrib of a pinnate frond in ferns; (2) the axis of a spike or raceme; rhachis.

RĀ′DIAL, pertaining to a radius, ray, or border; developed uniformly on all sides around a longitudinal axis. Compare DORSIVENTRAL.

RĀ′DĬAL BŬN′DLE, the axial fibrovascular system in roots and some lycopods—so called because it consists of alternating radial bands of xylem and phloëm.

RĀ′DĬANT, spreading from a common centre; radiating; radiate. Also having a ray or border, as the inflorescence of hydrangea.

RĀ′DĬĀTE, having rays or rayflorets.

RĀ′DĬĀTE-VEINED, see PALMATELY VEINED.

RĀ′DĬĀTĬNG, see RADIANT.

RĂD′ĬCAL, proceeding from the root, or base of the stem.

RĂD′ĬCANT, rooting—applied only where roots proceed from the stem above ground; radicating.

RĂD′ĬCĀTĔD, having a root or roots.

RĂD′ĬCĀTĬNG, see RADICANT.

RĂDĬCĀ′TION, (1) the arrangement or disposition of the roots in the soil, i.e., whether spreading near the surface or running vertically downward, etc.; (2) the arrangement of the roots and rootlets on their respective axes (rhizotaxis)—corresponding to Phyllotaxis, Caulotaxis, and Inflorescence; (3) the act of taking root. Compare RAMIFICATION.

RĂD′ĬÇĔL, a rootlet.

RÀDĬÇĬC′ŌLOŬS, growing upon the roots of other plants, as certain parasitic fungi.

RÀDĬÇĬFLŌ′ROŬS, having the flower-stalk arising from the crown of the root.

RADIC'IFORM, root-like.

RAD'ICLE, the portion of an embryo below the cotyledons, including the caulicle and nascent root; by some now restricted to the extreme point of the caulicle from which the root develops.

RAD'ICOSE, having roots unusually large or otherwise remarkable.

RADIC'ULA (obs.), see RADICLE.

RAD'ICULE (obs.), see ROOTLET.

RADIC'ULOSE, bearing rootlets or rhizoids, especially if numerons.

RA'DIUS (pl. **Ra'dii**), see RAY.

RA'DIX (pl. **Rad'ices**), see ROOT.

RAG (Hort.), the placenta and dissepiments (core and membranes) in the orange and other citrus fruits.

RA'MAL, see RAMEAL.

RA'MEAL, growing on or pertaining to branches; ramal; rameous.

RAM'ENT, sing., see RAMENTA.

RAMEN'TA (sing. **Ramen'tum**), scale-like hairs of various forms, especially when attached at an end or side like those on the petioles of ferns; raments. Also applied to other scales. Compare LEPIDES.

RAMENTA'CEOUS, bearing ramenta.

RAMEN'TUM, sing., see RAMENTA.

RA'MEOUS, see RAMEAL.

RA'MI, pl., see RAMUS.

RAMIF'EROUS, see RAMOSE.

RAMIFICA'TION, (1) the act or process of branching; (2) a branch, division, or offshoot; (3) the general arrangement of a system of branches. Applied to either stems or roots. Compare RADICATION.

RAMIFLO'ROUS, producing flowers along the larger branches.

RAM'IFORM, branch-like.

RAMIP'AROUS, producing branches.

RAMOSE', bearing branches; divided into branches; especially, divided into many branches; ramous.

RA'MOUS, see RAMOSE.

RAM'ULI, pl., see RAMULUS.

RAM'ULOSE, full of small branches.

RAM'ULOUS, see RAMULOSE.

RAM'ULUS (pl. **Ram'uli**), a small branch or twig.

RA'MUS (pl. **Ra'mi**), a branch.

RANK, a row, especially a vertical row of leaves or other organs upon a stem.

RA'PHE, the adherent funiculus connecting the hilum and chalaza in anatropous or amphitropous seeds. Also applied to a median line on the frustules of many diatoms, and formerly to the suture between the two halves of the fruit in Umbelliferæ; rhaphe.

RAPH'IDES, needle-shaped crystals; rhaphides. Sometimes improperly applied to plant-crystals of other forms. They are usually composed of oxalate of lime, and occur in large numbers in certain plants.

RAPHID'IAN, pertaining to raphides. A *raphidian cell* is a cell containing raphides.

RARE, sparse or few. Seldom used in this sense.

RARE'-RIPE, see RATH-RIPE.

RATH'-RIPE (rare), ripening or maturing early in the season; rare-ripe.

RATOON', a sprout from the root or base of a plant which has been cut off (applied mainly to sugar-cane). Compare SUCKER and PLANT-CANE.

RAUMPÄR'SĬT, see AULOPHYTE.

RĀY, a branch of an umbel or other somewhat radiating inflorescence; the ligulate corolla of an outer floret in a head of Compositæ; the margin of any circular surface, as distinguished from the central portion or disk.

RĀY-FLŌ'RĔT, see LIGULATE FLORET.

RĀY-FLOW'ĔR, one of the marginal florets of a head in Compositæ; ray-floret; ligulate floret. Also, a marginal flower in other flat-topped flower-clusters, especially when different from those of the centre or disk, as in hydrangea.

RĔÇĔP'TÂCLE, the place upon or within which anything is borne, as the summit of a flower-stalk upon which the floral organs are arranged, or upon which the flowers of a head are situated.

RĔÇĔPTĂC'ŪLAR, pertaining to or growing upon a receptacle.

RĔÇĔPTĂC'ŪLAR TŪBE, see CALYX-TUBE. Compare HYPANTHIUM.

RĔÇĔP'TĪVE SPŎT, the place in an oösphere at which the male gamete enters; i.e., at which fertilization takes place.

RĔÇĔSS', see SINUS.

RĔC'LĪNĀTE, bent over so that the apex is lower than the base, or until it rests on some support; reclined; reclining. Said of stems or branches when erect or ascending at the base, then turning toward the ground; of leaves in the bud when the blade is bent down upon the petiole, etc. When applied to a vine growing upon the ground it has about the same meaning as Recumbent and Decumbent.

RĔCLĪNED', see RECLINATE.

RĔCLĪN'ĪNG, see RECLINATE.

RĔCRUDĔS'ÇENÇE (obs.), see REJUVENESCENCE.

RĔC'TĪNĔRVED, parallel-veined.

RĔCTĬPĔTĂL'ĬTY̆, the tendency of growing organs to grow in a straight line. (Vochting.)

RĔCTĪSĒ'RĬAL, in straight rows.

RĔCÛR'VĀTE, see RECURVED.

RĔCŪRVED', curved outward or backward to a moderate extent, between forty-five and ninety degrees. Compare REFLEXED.

RĔDŪ'PLĬCĀTE, applied to parts of a flower in æstivation when they are valvate and have the margins reflexed.

RĔDŪPLĬCĀ'TION, an increase in the number of parts in a flower according to the plan upon which it is founded, i.e., by the addition of similar whorls; regular multiplication.

RĔFLĔCT'ĔD, see REFLEXED.

RĔFLĔXED', turned backward or outward more than Recurved, or to about the same extent but abruptly; reflected. Compare RECURVED and DEFLECTED.

RĔFLŌRĔS'ÇENÇE, blossoming anew.

RĔFRĂCT'ĔD, bent abruptly outward or backward, at an angle of more than forty-five degrees. Compare DEFLEXED and RECURVED.

RĔGĔNĔRĀ'TION, see REJUVENESCENCE.

RĒ'GION, an extent of country enclosing particular species, which are distributed through it in the stations adapted to their growth, and which, owing to natural conditions, are not generally found elsewhere, as the Arctic and Mediterranean regions.

RĔG'MȦ, a fruit with elastically dehiscing cocci, as in Euphorbia.

RĔG'MȦCÄRP, see REGMA.

RĔGRĔS'SION, see RETROGRESSIVE METAMORPHOSIS.

RĔG'ŪLAR, said of a flower which has all the parts of each whorl alike in size and form; actinomorphous.

RĔG'ŪLAR PĒLŌ'RĬȦ, when a flower, usually regular, becomes irregular by the non-development of some part. Compare PELORIA, IRREGULAR PELORIA, and EPANODY.

RĒJŪVĒNĔS'ÇENÇE, (1) the formation of a single new cell from the entire contents of a cell already formed ("renewal of cells"); especially, where the new cell escapes from the old cell-wall and develops a new plant, as in Œdogonium and some other algæ; (2) the renewal of growth and vigor as the result of conjugation or other form of fertilization (rare); (3) any renewed growth, or manifestation of increased vigor, as the annual growth of new canes in the raspberry or the formation of vigorous shoots from near the ground in other shrubs; renewal; regeneration.

RĒLĬQ'UĬÆ, see INDUVIÆ.

RĒMŌTE', separated by greater intervals than usual.

RĒNEW'AL OF ÇĔLLȘ, see REJUVENESCENCE.

RĔN'ĬFÔRM, kidney-shaped: heart-shaped, but broader than long and very deeply lobed at the base.

RĒPĂND', having a slightly undulating or sinuous margin. Compare SINUATE.

RĒ'PENT, see CREEPING.

RĔP'LĬCĀTE, folded backward at the sides or end. Compare REVOLUTE.

RĒ'PLŬM, a frame-like placenta in Cruciferæ and certain other plants across which the septum stretches, and from which the two valves fall away in dehiscence.

RĒPRŌDŬC'TĬVE ÔR'GANȘ, in flowering plants, the stamens and pistils.

RĔP'TANT, see CREEPING.

RĒȘĒRVE MȦTĒ'RĬAL, assimilated plant-food which is capable of again becoming food to serve in the growth of the plant, as starch.

RĔȘ'ĬN-ÇĔLL, a cell which has the office of secreting resin.

RĔȘ'ĬN-DŬCT, see RESIN-PASSAGE.

RĔȘ'ĬN-GLĂND, a cell or group of cells which secrete resin.

RĔȘĬNĬF'ĔROŬS, producing resin.

RĔȘ'ĬN-PĂS'SAĠE, an intercellular space or continuous gland in or into which resin is secreted; resin-duct; resin-tube.

RĔȘ'ĬN-TŪBE, see RESIN-PASSAGE.

RĔSPĬRĀ'TION, in a broad sense, includes all transfers of gases and vapors between the plant and the air. As usually employed it refers to the absorption of carbon dioxide from the air, and its decomposition in the plant, with the assimilation of the carbon and the liberation of the oxygen. In a strict sense it applies to the same process as respiration in animals, though existing in plants to a feebler degree, namely, the reception of oxygen, the oxidation of oxidizable matter, and the release of the products of oxidation. See TRANSPIRATION.

RĒSPĪ'RÅTŌRY̌ CǍV'ǏTY̌, see STOMATIC CHAMBER.

RĒSPĪ'RÅTŌRY̌ CHĀM'BĒR, see STOMATIC CHAMBER.

RĔS'TANT, see PERSISTENT.

RĔST'ǏNG ÇĔLL, see RESTING-SPORE.

RĔST'ǏNG-NŪ'CLĒŬS, a nucleus when not in the act of karyokinesis. The nuclear filament then exists in the form of granules, or as a network, often with one or more nucleoli.

RĔST'ǏNG PĒ'RǏOD, the period during which a plant, seed, or spore remains dormant. Applied mainly to the dormant period of a resting-spore, then also called by some "involution-period."

RĔST'ǏNG-SPŌRĂN'ĠǏŬM, a term applied by Pringsheim to certain dormant gonidia of Saproleguia and related fungi which eventually produce swarm-spores.

RĔST'ǏNG-SPŌRE, a spore furnished with thick walls and able to retain its vitality for a considerable time, and generally requiring the lapse of time, often over winter, before it will germinate; winter-spore; hypnospore; statospore.

RĔST'ǏNG-STĀĠE, the stage or condition of a plant, bulb, seed, or spore during its dormant or resting period.

RĒSŪ'PǏNĀTE, upside-down; inverted; retroverted; reversed. Compare SUPINE.

RĒTǏC'ŪLĀTE, in the form of network, as the veins in the leaves of most plants.

RĒTǏC'ŪLĀTĔD, see RETICULATE.

RĒTǏC'ŪLĀTĔD VĔS'SĔL, one having thickenings in the form of network.

RĒTǏCŪLĀ'TION, any network; reticulum.

RĒTǏC'ŪLŬM, a fibrous membrane at the base of the petioles in palms (mattulla); the network which pervades the substance of the cell and nucleus enclosing the softer portions of the protoplasm; any system of filaments or fibres in the form of network.

RĔT'ǏFÔRM, see RETICULATE.

RĔTǏNĂC'ŪLŬM (pl. **Rĕtĭnăc'ūlå**), (1) the viscid glandular disk derived from the rostellum at the base of the caudicle of a pollinium in orchids by which it is attached to a visiting insect (cement-disk of Müller); (2) one of the arms connecting the pollinia of milk-weeds with the corpusculum; (3) the persistent indurated hook-like funiculus of the seeds in most Acanthaceæ.

RĔT'ǏNĔRVED, see RETICULATE-VEINED.

RĒ'TRŌFLĔXED, see REFLEXED.

RĒ'TRŌFRĂCTĔD, see REFRACTED.

RĒTRŌGRĔS'SION, see RETROGRESSIVE METAMORPHOSIS.

RĒTRŌGRĔS'SǏVE MĔTÅMÔR'PHŌSǏS, the occurrence in the place of organs of the usual character of those belonging to a lower condition or position, as when pistils become converted into stamens, petals, or leaves; descending metamorphosis; retrogression; regression. Compare PROGRESSIVE METAMORPHOSIS.

RĔTRÔRSE′, turned backward in any manner.

RĔTRÔRSE′LY̆ ĂCŪ′LEĀTE, having prickles directed backward or downward, as on the stem of *Galium asprellum.*

RĒ′TRŌVĔRTĔD, see RESUPINATE.

RĔTŪSE′, slightly notched at a rounded apex. Compare EMARGINATE.

RĔVĔRSED′, upside-down; resupinate; directed backward toward the base; extending in an opposite to the usual direction.

RĔVĔR′SION, appearance in an earlier, lower, or simpler form than usual. See RETROGRESSIVE METAMORPHOSIS and ATAVISM.

RĔV′ŌLŪTE, rolled outward, backward, or downward. Compare INVOLUTE and CIRCINATE.

RĒVŎLV′ĬNG NŪTĀ′TION, see CIRCUMNUTATION.

RHĂB′DŌID, an acicular or spindle-shaped protoplasmic body found in the stalk-cells of the tentacles of Drosera and in the mesophyll cells of Dionæa; plastoid. It tends to become spherical when the part is stimulated, but its full significance is unknown.

RHĀ′CHĬS, see RACHIS.

RHĀ′PHĒ, see RAPHE.

RHĂPH′ĬDĒS, see RAPHIDES.

RHĒŎT′RŌPISM, having the direction of growth determined (not mechanically) by a current of water. (Jönsson.) See POSITIVELY and NEGATIVELY RHEOTROPIC. Compare HELIOTROPISM.

RHĪZĂMÔR′PHŌID, see RHIZOMORPHOUS.

RHĪZĂN′THOŬS, parasitic upon the roots of another plant, and producing flowers with little or no green foliage of its own, as Rafflesia and Monotropa. A term of little importance, seldom used. Compare RADICIFLOROUS.

RHĬZĪ′NĀ (pl. **Rhĭzī′næ**), see RHIZOID.

RHĪ′ZĪNE, see RHIZOID.

RHĪZŌCĂR′POŬS, having a perennial root and herbaceous stem; rhizocarpic. (Rare.)

RHĬZ′ŌGĔN, (1) a plant which is parasitic upon the roots of another; (2) any part or organ which emits roots or rhizoids.

RHĪZŌGĔNĔT′ĬC, root-producing, as *rhizogenetic* tissue; rhizogenic.

RHĬZŌGĔN′ĬC, see RHIZOGENETIC.

RHĪ′ZŌID, adj., see RHIZOMORPHOUS.

RHĪ′ZOID, n., a root-like organ in many cryptogams; rhizina; rhizine. Compare ROOT.

RHĪZŌ′MĀ (pl. **Rhīzō′mātā**), see RHIZOME.

RHĬZŌMĂT′ĬC, having the appearance or character of a rhizome.

RHĬZŌME′, a subterranean stem, especially if uniformly thickened for the storage of starch, etc.; rhizoma; root-stock. Compare STOLON and TUBER.

RHĪ′ZŌMÔRPHS, root-like organs in Agarics and some other fungi composed of many united hyphal strands.

RHĪZŌMÔR′PHOŬS, root-like; rhizomorphoid; rhizoid.

RHĪ′ZŌPHŌRE, a structure in certain species of Selaginella which resembles a root and from which true roots are developed.

RHĪZŌPHȲL'LOŬS, emitting roots from the leaves. (Rare.)

RHĬZŌTĂX'ĬS, the position or order of arrangement of the roots upon a plant; radication. Compare PHYLLOTAXIS and CAULOTAXIS.

RHĪ'ZŌTĂXȲ, see RHIZOTAXIS.

RHŌDŎL'ŌĠȲ, the part of botany which treats of roses.

RHŌ'DŌPHȲL, the compound pigment which is the coloring matter of red algæ, the Florideæ.

RHŌDŌSPĔR'MĬN, crystalloids of proteid found in the Florideæ containing the rhodophyl or red coloring matter.

RĬB, a large prominent vein.

RĬBBED, (1) having more than one prominent longitudinal vein; (2) applied to an apple or other fruit with large longitudinal ridges.

RĬC'TŬS, an old term for the throat of a personate flower.

RĬĠĔS'ÇĔNT, approaching a rigid or stiff consistence. (M. C. Cooke.)

RĬGHT, see remarks under DEXTRORSE and LEFT.

RĪ'MȦ (pl. **Rī'mæ**), a chink or cleft.

RĪMŌSE', abounding with chinks, clefts, or cracks, as the bark of many trees and the thallus of certain lichens; rimous.

RĪM'OŬS, see RIMOSE.

RĬM'ŪLŌSE, diminutive of Rimose.

RĬND, see CORTEX.

RĬNG, see ANNULUS.

RĬNGED, see ANNULATED.

RĬN'ĠENT, bilabiate, with the lips widely separated and the throat open, as in Lamium. Compare PERSONATE.

RĬPĀ'RĬAN, see RIPARIOUS.

RĬPĀ'RĬOŬS, growing along rivers or other water-courses; riparian.

RĬVŌSE', having sinuate channels. Compare SULCATE.

RĬV'ŪLŌSE, having small sinuate channels. Compare STRIATE.

RŎD - FRŬCTĬFĬCĀ'TION, unbranched gonidiophores in Basidiomycetes.

RŌGUE (Hort.), an inferior sport or variation, i.e., a plant which deviates in an undesirable manner from the type of the species or variety.

RO̅O̅T, the descending axis;—differing in character from the stem chiefly in being destitute of leaves. Compare RHIZOID and RHIZOME. Sachs extends the term Root to the root-like organs (rhizoids) of cellular plants.

RO̅O̅T-CĂP, a protective covering to the growing extremity of roots; calyptra; pileorhiza.

RO̅O̅T-HÂIRS, slender hairs upon the younger roots, which serve to absorb food from the soil.

RO̅O̅T - LĒAF, a leaf springing from the base of the stem.

RO̅O̅T - PRĔS'SŪRE, see SAP-PRESSURE.

RO̅O̅T'LĔT, a small root; root-fibre.

RO̅O̅T-SHĒATH, see COLEORHIZA.

RO̅O̅T'STŎCK, see RHIZOME.

RŌ'RĬDŬS, covered with transparent elevations resembling drops of dew. (Obs.)

RŌSĀ'ÇEOŬS, rose-like, or belonging to the family Rosaceæ.

RŌSĔL'LȦ (obs.), see ROSETTE.

RŌSĔTTE', a cluster of leaves or other organs arranged somewhat like the petals of a double rose, as the leaves of the dandelion or those upon the short spurs of the larch.

RŌS′TĔL, see ROSTELLUM.

RŌS′TĔLLĀTE, diminutive of Rostrate, having a small beak.

RŌSTĔL′LĬFŌRM, beak-shaped, especially when the process is small; rostriform.

RŎSTĔL′LŬM (pl. **Rŏstĕl′là**), a little beak; especially, the process upon the column in orchids containing the disk (retinaculum) of the pollinia. It is one of the three united styles which has become modified for this purpose.

RŌS′TRĀTE, see BEAKED.

RŌS′TRĬFŌRM, beak-shaped.

RŌS′TRŬM, see BEAK.

RŌ′ȘŪLÀ (obs.), see ROSETTE.

RŌ′ȘŪLAR (obs.), see ROSULATE.

RŌ′ȘŪLĀTE, arranged or shaped like a rosette; having the leaves in little tufts like the petals of a double rose, as those of the dandelion.

RŌ′TĀTE, wheel-shaped. In flowers, applied to a gamopetalous corolla with a very short tube and nearly flat spreading border.

RŌ′TĀTE-PLĀNE, gamopetalous, with a flat border and no tube.

RŌTĀ′TION, a flowing movement of the protoplasmic cell-fluid; cyclosis; streaming. In some active or growing cells the movement is readily visible as distinct currents having a rotary or back-and-forth direction. The term Cyclosis is now well established, and is perhaps preferable to Rotation as a general term, the latter word being applied especially to circulation next the cell-wall, to a rotation of the whole mass within the cell, or to a peculiar spiral movement of the cell-fluid in certain plants, as in Chara.

RŌ′TĬFŌRM (obs.), see ROTATE.

RŌTŬND′, rounded; somewhat orbicular.

ROŬGH, see SCABROUS.

ROŬGH LĒAVEȘ, a gardener's term for the true leaves which succeed the cotyledons.

RU̥BĔS′ÇENT, reddish or rosy; rubicund.

RU̥′BĬCŬND, see RUBESCENT.

RU̥BĬĠ′ĬNŌSE, of the color of iron-rust; brownish red; rubiginous; rufous; rusty; ferruginous.

RU̥BĬĠ′ĬNOŬS, see RUBIGINOSE.

RU̥′DĔRAL, growing in rubbish or waste places.

RU̥′DĬMENT, an imperfectly developed, vestigal, or aborted part.

RU̥DĬMĔN′TAL, see RUDIMENTARY.

RU̥DĬMĔN′TÀRY̆, imperfectly developed, or in an early stage of development; rudimental.

RU̥FĔS′ÇENT, somewhat rufous.

RŬF′FLED, sometimes applied to a margin when very strongly waved.

RU̥′FOŬS, see RUBIGINOSE.

RU̥′GÀ, a wrinkle.

RŬG′GĔD (obs.), see SCABROUS.

RU̥GŌSE′, wrinkled, as the leaves of sage. Compare BULLATE and CRISP.

RU̥′GŪLŌSE, slightly rugose.

RU̥′MĬNĀTĔD, said of the albumen or cotyledons of a seed which has holes or channels into which the inner seed-coat penetrates, as in the papaw and nutmeg.

RŬN'ÇĬNĀTE, having large sharp teeth or lobes projecting backward, as the leaves of dandelion.

RŬN'NĔR, a stolon which roots and forms new plants at intervals, as the strawberry; the prostrate stem of a vine, as in melons.

RŬN'NĬNG, prostrate upon or beneath the surface, but not rooting. Often used improperly in the sense of Creeping, which see.

RUPĔS'TRĬNE, growing upon rocks or in rocky places; lapidose; saxatile; saxicole; saxicoline; saxicolous; rupicoline; rupicolous. The distinctions between these terms are mainly slight and inconstant. Lapidose may, however, be defined as growing upon land abounding in loose stones or coarse gravel (compare GLAREOSE), while the other terms usually refer to fixed or massive rock. There is a tendency, also, to restrict the terms Rupicolous and Saxicolous to growth directly upon the rocks, as with many lichens, mosses, and some ferns.

RUPĬC'ŌLOUS, see RUPESTRINE.

RŬP'TĬLE, dehiscing in an irregular or accidental manner.

RŬST'Y̆, see RUBIGINOSE.

SĂB'ŪLĬNE, see ARENACEOUS.

SĂB'ŪLŌSE, see ARENACEOUS.

SĂC, a deep or closed receptacle. Also written Sack.

SĂC'CĀTE, in the form of a bag or pouch; sacciform.

SĂC OF THE ĂM'NĬŎS (obs.), see EMBRYO-SAC.

SĂC OF THE ĔM'BRY̆Ō, see EMBRYO-SAC.

SĂCK, see SAC.

SĂG'ĬTTĀTE, arrow-shaped. Differs from Cordate in having the lobes as well as apex acute.

SĀ'LĬENT, projecting outward.

SĀ'LĬNE, growing upon the seashore or other places impregnated with salt; salsuginous.

SĂLSŪ'ĠĬNOŬS, see SALINE.

SĂL'VĔR-FŌRM, see HYPOCRATERIFORM.

SĂL'VĔR-SHĀPED, see HYPOCRATERIFORM.

SĂMĀ'RĂ, a winged fruit, like that of the ash, elm, or maple; pteridium; key.

SĂM'ĂROĪD, winged like a samara.

SĂP, in general, any vegetable fluid, especially limpid fluid, such as abounds in many trees in spring. See CELL-SAP.

SĂP-CĂV'ĬTY̆, see VACUOLE.

SĂP'ĬD, having a pleasant taste.

SĂP'LĬNG, a tall seedling tree having a stem from two to six inches in diameter.

SĂP-PRĔS'SŪRE, the force exerted by the soil-fluid in entering the plant, and by the sap in passing upward within it; root-pressure. See OSMOSE.

SĂPRŎĠ'ĔNOŬS, causing putrefaction.

SĂP'RŌPHY̆TE, a plant which lives upon dead animal or vegetable matter; humus-plant. Compare PARASITE.

SĂPRŌPHY̆T'ĬC, living upon dead organic matter. Compare PARASITIC.

SĂP-VĔS'SĔL, see VESSEL.

SĂP'-WOOD, see ALBURNUM.

SÄRCŌB'ĂSĬS, see CARCERULE.

SÄR'CŌCÄRP, the succulent fleshy portion of a fruit, especially of a drupe. Used by

M. T. Masters synonymously with Berry.

SÄR'CŌDĔRM, a fleshy layer in certain seed-coats; sarcoderma.

SÄR'MĔNT, a long slender runner, stolon, or twig; sarmentum; flagellum.

SÄRMĔNTĀ'ÇEOŬS, see SARMENTOSE.

SÄRMĔNTĬF'ĔROŬS, bearing sarments.

SÄRMĔNTŌSE', bearing or resembling sarments; sarmentaceous; sarmentous.

SÄRMĔN'TOŬS, see SARMENTOSE.

SÄRMĔN'TŬM (pl. **Särmĕn'tȧ**), see SARMENT.

SẠU'SĀĠE-SHĀPED, see ALLANTOID.

SẠW-TOOTHED, see SERRATE.

SĂX'ĀTĬLE, see RUPESTRINE.

SĂX'ĬCŌLE, see RUPESTRINE.

SĂXĬC'ŌLĬNE, see RUPESTRINE.

SĂXĬC'ŌLOŬS, growing upon rocks, as many mosses and lichens; rupicolous. Compare RUPESTRINE.

SCĀ'BRĀTE, see SCABROUS.

SCĀ'BRĬD, slightly scabrous.

SCĀ'BRĬDOŬS, somewhat scabrid or scabrous.

SCȦBRĬD'ŪLOŬS, slightly scabrid or scabrous; scabridous.

SCĀ'BROŬS, rough with short stiff hairs or points; scabrate; asperate.

SCȦLĂR'ĬFŌRM, having transverse bars like a ladder: applied mainly to thickenings of this form characteristic of scalariform vessels.

SCȦLĂR'ĬFÔRM DŬCT, see SCALARIFORM VESSEL.

SCȦLĂR'ĬFŌRM MÄRK'ĬNG, an elongated pit of a scalariform vessel.

SCȦLĂR'ĬFÔRM VĔS'SĔL, a vessel marked with elongated transverse bordered pits, as in many ferns.

SCĀLE, any thin, scale-like appendage, usually a degenerated and metamorphosed leaf, as those of buds and buds. See RAMENTUM and LEPIDE.

SCĂL'LION, a small or imperfect onion, particularly the shallot, *Allium Ascalonicum*. Compare SCULLION.

SCẠL'LȮPED, see CRENATE.

SCĀL'Ȳ, consisting of scales; furnished or covered with scales; scale-like in texture.

SCĀL'Ȳ BŬLB, one with narrow and somewhat separate scales, as in the lily. Compare TUNICATED BULB.

SCĂN'DENT, see CLIMBING.

SCĀPE, a peduncle rising from the ground, as in blood-root, *Sanguinaria Canadensis*; i.e., a stalk from the root or collar which bears one or more flowers but no foliage-leaves.

SCĂPH'OID, see NAVICULAR.

SCĀPĬFLŌ'ROŬS, having the flowers on scapes.

SCĀ'PĬFŌRM, scape-like; scapoid.

SCĀPĬĠ'ĔROŬS, bearing scapes.

SCĀ'POID, see SCAPIFORM.

SCĀ'PŬS, see SCAPE.

SCÄR, the mark left by the natural separation of a leaf or other organ; cicatrix. See LEAF-SCAR and HILUM.

SCĀ'RĬŌSE, see SCARIOUS.

SCĀ'RĬOŬS, dry and membranous; scariose.

SCĂT'TĔRED, either sparse, or without apparent regularity of arrangement. See ALTERNATE.

SCHIZ'OCARP, a fruit of which the carpels separate when mature and retain the seeds in cocci or mericarps.

SCHIZOGENET'IC, see SCHIZOGENIC

SCHIZOGEN'IC, formed by the separation of parts, as some intercellular spaces; schizogenetic; schizogenous. Compare LYSIGENIC.

SCHIZOG'ENOUS, see SCHIZOGENIC.

SCIM'ETAR-SHAPED, see ACINACIFORM.

SCI'ON, formerly used for any shoot or twig; now applied only to one intended for grafting. Improperly written Cion. A scion is considered by most nurserymen to be a part of a twig or shoot long enough to make one graft, usually four to six inches.

SCI'UROID, in the shape of a squirrel's tail.

SCLEREN'CHYMA, lignified tissue of any kind, especially such as composes the shell of a nut. It is not usually applied to ordinary woody tissue, but is used by many botanists for thick-walled fibres of the bast and hypoderma.

SCLER'OGEN, see LIGNINE.

SCLE'ROID, woody or bony in texture.

SCLERO'SIS, see LIGNIFICATION.

SCLE'ROTE, see SCLEROTIUM.

SCLEROT'IC, resembling or consisting of sclerenchyma.

SCLERO'TIUM (pl. **Sclero'tia**), a consolidated and hardened mass of hyphæ in a resting condition, as in Claviceps (ergot) and some other Ascomycetes. Also a temporary condition assumed by the plasmodium of Myxomycetes, chiefly due to dryness, in which it appears in dry, compact, waxlike, gritty nodules.

SCLEROT'OID, consisting of or resembling a sclerotium.

SCOBIC'ULAR, see SCOBIFORM.

SCOB'IFORM, resembling sawdust; scobicular.

SCOBI'NA, an old term for the zigzag rachis of certain grasses.

SCOL'ECITE, Tulasne's later term for the "vermiform body" of Woronin. It is the ascogonium or archicarp in certain Ascomycetes, as *Ascobolus pulcherrimus*.

SCOR'PIOID, commonly used in the sense of Helicoid, but properly restricted now to alternately progressive or zigzag development. See SCORPIOID CYME and SCORPIOID DICHOTOMY.

SCOR'PIOID CYME, one in which the successive flowers are situated on alternate sides of the pseudaxis; cicinal cyme; cicinus or cincinnus. The term Scorpioid Cyme was given by A. P. de Candolle to a unilateral cyme, the undeveloped portion of which is usually rolled up like the tail of a scorpion, as in Boraginaceæ, and this is the meaning still attached to the term by most botanists of England. Compare HELICOID CYME.

SCOR'PIOID DICHOT'OMY, one in which alternate branches develop in the successive bifurcations; cicinal dichotomy. Compare HELICOID DICHOTOMY.

SCROBIC'ULATE, pitted.

SCRŌ'TĬFÔRM, see POUCH-SHAPED.

SCŬL'LION, a term applied by gardeners in America to onions which fail to "bottom" properly, but remain soft and become thick-necked. Probably derived from Scallion, which see.

SCŬLP'TŪRED, a horticultural term applied to the seeds of certain gourds, etc., which have peculiar markings.

SCÛRF, small bran-like epidermal scales.

SCŪ'TĀTE, shield-shaped, especially like a round shield; scutiform; clypeate. Compare PELTATE.

SCŪ'TĔL, see SCUTELLUM.

SCŪTĔL'LÆFÔRM, see SCUTELLIFORM.

SCŪ'TĔLLĀTE, diminutive of Scutate; somewhat scutate, like a platter; especially, covered with small plate- or shield-like surfaces; scutellated. See SCUTELLIFORM.

SCŪ'TĔLLĀTĔD, see SCUTELLATE.

SCŪTĔL'LĬFÔRM, shaped like a saucer or small platter; scutellate; scutellæform. Compare PATELLIFORM.

SCŪTĔL'LŬM (pl. **Scūtĕl'là**), (1) a sessile apothecium or "shield" in lichens; (2) in Gramineæ a usually shield-like expansion of the hypocotyl which acts as an organ of suction through which the embryo absorbs nutriment from the endosperm. It is regarded as the cotyledon. It appears exteriorly as a large flat space upon the surface of the seed surrounding the embryo (plumule and radicle) to which it is attached. It forms the chief part of what is known as the "chit" in a kernel of corn. Compare SCUTUM.

SCŪ'TĬFÔRM, see SCUTATE.

SCŪ'TĬFÔRM LĔAF, the first-formed leaf (protophyl) in Salvinia, differing in form from the succeeding leaves.

SCŪ'TŬM, a large circular disk-like part or organ, as the dilated stigma (stylostegium) of Stapelia; shield. Compare SCUTELLUM.

SÇY̆M'ĬTAR-SHĀPED, see ACINACIFORM.

SÇȲ'PHĬFÔRM, like a scyphus; goblet-shaped.

SCȲ'PHŬS (pl. **Scȳ'phī**), an apothecium in lichens elevated on a podetium. Compare SCUTELLUM.

SĒBĀ'ÇEOŬS, wax-like, or producing wax.

SĒBĬF'ĔROŬS, producing wax.

SĔC'ŌNDĀRY̆, opposed to Primary; often includes Tertiary, etc.

SĔC'ȮNDĀRY̆ BŬDS, see ACCESSORY BUDS.

SĔC'ŌNDĀRY̆ CŌR'TĔX, a collective term for the successive formations of liber or bast within the cortical sheath and primary cortex. It is not applied to secondary cork formations.

SĔC'ȮNDĀRY̆ FŬN'GŬS, a saprophytic or parasitic fungus which attacks a plant after it has been killed or injured by another fungus. Compare SUPERPARASITE.

SĔC'ȮNDĀRY̆ HȲ'BRĬD, see DERIVATIVE HYBRID.

SĔC'ȮNDĀRY̆ MĔD'ŬLLĀRY̆ RAYS, those intermediate between the rays first formed, and which do not extend to the pith.

SĔC'ŌNDĀRY̆ MĔR'ĬSTĔM, meristem in an organ or part after its first development is completed and by means of which further growth is effected, as cambium. Compare PRIMARY MERISTEM.

SĔC'ŌNDĀRY̆ NŪ'CLĒŬS, the "nucleus of the embryo-sac," resulting from the union of the two polar nuclei.

SĔC'ONDĀRY̆ PĒDŬN'CLE, the stalk of a branch of an inflorescence bearing more than one flower; partial peduncle. Compare PEDICEL.

SĔC'ŌNDĀRY̆ PĔT'ĬŌLE, the stalk of a leaflet; partial petiole petiolule.

SĔC'ŌNDĀRY̆ ROOT, a lateral root, especially one growing from the stem. Compare PRIMARY ROOT and AËRIAL ROOT.

SĔC'ŌNDĀRY̆ SPŌRE, a spore borne on a promycelium (sporidium), or one derived immediately from another spore.

SĔC'ŌNDĀRY̆ STRŬC'TŪRE, the structure or condition of a plant or organ after its first year, or after it has grown beyond its first-formed state. Compare PRIMARY STRUCTURE.

SĔC'ŌNDĀRY̆ WOOD, the portion of the wood or xylem which is formed after the primary xylem, or after the formation of the cambium ring. Compare PRIMARY WOOD.

SĔC'ŌNDĪNE (obs.), see SECUNDINE.

SĔC'TĬLE, divided into separable parts.

SĒ'CŬND, either unilateral (one-ranked) or homomalous (one-sided), i.e., having lateral organs upon one side only, or all turned to one side.

SĔC'ŬNDĪNE, the secondary or inner coat of an ovule, called tegmen in the seed. By some authors this inner coat, since it is first formed, is called the Primine and the outer coat the Secundine

SEED, a fertilized ovule: a structure enclosing when mature a rudimentary plant (the embryo) which is protected while in a dormant condition, and furnished with nutriment, either in its cotyledons or around them, with which to begin the process of growth.

SEED'ĀGE, a term proposed by L. H. Bailey for the process, state, or condition of being propagated by seed.

SEED-BŬD, an old term used both for ovule and ovary.

SEED-CŌAT, the covering of a seed. See TEGMEN and TESTA.

SEED-LEAF, see COTYLEDON.

SEED'LĬNG, (1) a plant produced from seed, in distinction from one produced in any other manner, either naturally or artificially; (2) a young plant produced from seed, in distinction from one of considerable age or size.

SEED-LŌBE, see COTYLEDON.

SEED-SPŌRT, a sport which originated from seed; seminal sport. Compare BUD-SPORT.

SEED-VĀRĬĀ'TION, a variation derived from seed instead of buds. Compare BUD-VARIATION.

SEED-VĀRĪ'ĔTY̆, any variety which originated from seed, either suddenly (seed-variation or seed-sport) or gradually in the ordinary way, and is propagated in any manner. Compare BUD-VARIETY.

SEED-VĔS'SĔL, see PERICARP.

SĔG'MENT, a natural division or part; as one of the parts of a compound or deeply divided leaf, an internode, the cell of a pericarp, any joint or regular part whether separable or not.

SĔGMĔNTĀ'TION, the act or process of dividing into segments.

SĔG'RĔGĀTĔD, separated.

SEĬ'RŌSPŌRE, a kind of gonidium produced in mouilliform rows in certain red algæ.

SĔLF, a florist's term for a flower or plant which is wholly of one color.

SĔLF-CŌL'ŌRED, applied to fruits and flowers which are of one color; unicolor. Compare COLORED.

SĔLF-FĔRTĬLĬZĀ'TION, see CLOSE-FERTILIZATION.

SĔMĬ-, a Latin prefix meaning half or partial.

SĔMĬĂDHĔR'ENT, adherent halfway, as the calyx half-way to the summit of the ovary.

SĔMĬĂMPLĔX'ĬCAUL, slightly clasping the stem.

SĔMĬĂNĂT'RŌPOŬS, see AMPHITROPOUS.

SĔMĬCŌLŬM'NAR, see SEMITERETE.

SĔMĬCŌR'DĀTE, in the shape of a lateral half of a cordate body.

SĔM'ĬDOŬBLE, when a part only of the stamens are replaced by petals. Compare FULL.

SĔMĬĔQ'UĬTANT, see HALF-EQUITANT.

SĔM'ĬFLŌRĔT, a floret in Compositæ having a strap-shaped corolla; ligulate floret; semifloscule.

SĔMĬFLŎS'CŪLAR, having the florets of a head in Compositæ all ligulate; semiflosculous.

SĔMĬFLŎS'CŪLE, see SEMIFLORET.

SĔMĬFLŎS'CŪLOŬS, see SEMIFLOSCULAR.

SĔMĬLĔNTĬC'ŪLAR, see SUBLENTICULAR.

SĔMĬLŎC'ŪLAR, having the dissepiments incomplete and the pericarp therefore really unilocular; semiseptate.

SĔMĬLŪ'NAR, see LUNAR.

SĔMĬLŪ'NĀTE, see LUNAR.

SĔM'ĬNAL, pertaining to seed.

SĔM'ĬNAL LĔAF, see COTYLEDON.

SĔM'ĬNAL SPŌRT, see SEED-SPORT.

SĔMĬNĀ'TION, the production of seeds. Sometimes used for their dissemination or dispersion.

SĔMĬNĬF'ĔROŬS, seed-bearing.

SĔMĬÔRBĬC'ŪLAR, in the shape of one half of a circular body. Compare SUBORBICULAR.

SĔMĬŌ'VAL, having the shape of half of an oval, divided lengthwise.

SĔMĬŌ'VĀTE, in the form of half of an ovate figure, divided lengthwise, as where one side of an otherwise ovate leaf is wanting.

SĔMĬRĀ'DĬĀTE, having a part only of the marginal florets of a head ligulate.

SĔMĬRĔN'ĬFŌRM, reniform, with one lobe wanting.

SĔMĬSĂG'ĬTTĀTE, sagittate, with one lobe wanting.

SĔMĬSĔP'TĀTE, see SEMILOCULAR.

SĔMĬSTĂM'ĬNĀTE, formerly sometimes used for semidouble.

SĔMĬTĔRĔTE', half-terete; half-cylindrical; semicolumnar.

SĔMĬVĂL'VĀTE, applied to a pericarp in which the valves are only partially dehiscent.

SĔMĬVĔRTĬÇ'ĬLLĀTE, see SUBVERTICILLATE.

SĔMPĔRVĪ'RENT, see EVERGREEN.

SĔN'SĬTĬVENĔSS, see IRRITABILITY.

SĔ'PAL, a leaf or lobe of a calyx.

SĔP'ALĬNE, pertaining to or resembling a sepal.

SĔPĂL'ŌDY̆, the reversion of petals to sepals.

SĔP'ALŌĬD, sepal-like.

SĔP'ALOŬS, having or relating to sepals; sepaline.

SĔP'ÅRĀTE, see FREE and DISTINCT.

SĔP'ÅRĀTĔD FLOW'ĔRS, those having stamens or pistils but not both.

SĔP'ÅRĀTĬNG LĀY'ĔR, see ABSCISS LAYER.

SĔP'TÅ, pl., see SEPTUM.

SĔPTĂM'ĔROŬS, having the parts in sevens.

SĔP'TĀTE, having one or more partitions (septa). Compare SEPTIFEROUS.

SĔP'TĀTE SPŌRE, see COMPOUND SPORE.

SĔP'TĔNĀTE, having seven parts, or the parts in sevens; septamerous.

SĔPTĬCĪ'DAL, a mode of dehiscence in which the separation takes place through the partitions, i.e , along the line of junction of the carpels. Compare SEPTIFRAGAL and LOCULICIDAL.

SĔPTĬF'ĔROŬS, bearing one or more partitions or septa—said of the valves of certain capsules after dehiscence. Compare SEPTATE.

SĔP'TĬFŌRM, resembling a septum.

SĔPTĬF'RÅGAL, a mode of dehiscence in which the valves break away from the septa or partitions. Compare SEPTICIDAL and LOCULICIDAL.

SĔP'TŪLĀTE, imperfectly or rarely septate, or having false or supernumerary septa.

SĔP'TŪLŬM (pl. **Sĕp'tūlȧ**), diminutive of Septum: a little septum or partition. Also a spurious or supernumerary septum.

SĔP'TŬM (pl. **Sĕp'tȧ**), a partition of any kind, as between two cells in a tissue, or between cavities in an ovary. See DISSEPIMENT.

SĔ'RĬAL, in rows or regular succession.

SĔ'RĬĀTE, see SERIAL.

SĔRĬÇ'EOŬS, silky; covered with fine, straight, soft, smooth, glossy, appressed hairs.

SĔ'RIĔS, (1) a row or rank; (2) a term applied to various subordinate groups of different values, and also used by Asa Gray as equivalent to Subkingdom, which see.

SĔRŎT'ĬNOŬS, occurring late, or comparatively late, in the season.

SĔR'RĀTE, having sharp marginal teeth projecting toward the apex.

SĔR'RĀTE-ÇĬL'ĬĀTE, having the teeth of a serrate margin tipped with fine slender hairs.

SĔR'RĀTĔD, see SERRATE.

SĔR'RÅTŪRE, one of the teeth of a serrate margin. Compare SERRULATION.

SĔR'RĬED, close together in rows, like the kernels in the rows upon an ear of Indian corn.

SĔR'RULĀTE, diminutive of Serrate; serrate with small teeth. Compare SUBSERRATE.

SĔR'RULĀTĔD, see SERRULATE.

SĔRRULĀ'TION, (1) the state or condition of being serrulate; (2) a tooth of a serrulate margin. Compare SERRATURE.

SĔS'QUĬ-ĂL'TĔR, half as many more: applied to stamens when half as many more than the petals; also to a fertile floret in grasses when accompanied by a small abortive one.

SĔS'SĬLE, (1) having no stalk, as a leaf which has no petiole but is seated directly upon the stem; (2) quiescent, not moving about --applied to bacteria and zoöspores at certain times.

SĔ'TÀ (pl. **Sē'tæ**), a bristle or slender bristle-like body; the stalk of the capsule in most mosses above the true stem.

SĒTĀ'ÇEOŬS, (1) bristle-shaped; setiform; (2) setigerous.

SĒTĬF'ĔROŬS, see SETIGEROUS.

SĔ'TĬFŌRM, see SETACEOUS.

SĒTĬĠ'ĔROŬS, bristle - bearing; setiferous. See SETOSE.

SĔTŌSE', bearing or abounding with bristles; bristly; setaceous; setous; setiferous; setigerous.

SĔ'TOŬS, see SETOSE.

SĔT'ŪLÀ (pl. **Sĕt'ūlæ**), diminutive of Seta.

SĔT'ŪLŌSE, bearing minute bristles.

SEX-, Latin for six. See HEX-.

SĔX, one of the attributes of nearly all living bodies, which manifests itself in a certain method of reproduction, the first stage of which is the blending of the contents of two cells which are usually of distinct character and different origin, one of which is called male, the other female.

SĔXĂṈ'GŪLAR, having six angles; hexagonal.

SĔXFĀ'RĬOŬS, six-rowed.

SĔX'ĬFĬD, six-cleft.

SĔXLŎC'ŪLAR, having six cells in an ovary.

SĔX'PÄRTĬTE, six-parted.

SĔX'ŪAL ĠĔNĔRĀ'TION, the generation or stage which bears the sexual organs in plants which have an alternation of generations. In ferns it is the prothallus. Compare ASEXUAL GENERATION.

SĔX'ŪAL SȲS'TĔM, see LINNÆAN SYSTEM.

SHĂG'GȲ, either villose or hirsute.

SHĔATH, the petiole or base of the leaf in grasses which surrounds the stem; any tubular part surrounding another; vagina.

SHĔATHED, surrounded by a sheath; invaginated; vaginate.

SHĔATH'ĬNG, surrounding a stem or other body in a convolute or tubular manner, as the petioles of grasses; vaginant. Compare AMPLEXICAUL.

SHĔLL, the bony covering of a nut.

SHIĔLD, see APOTHECIUM and INDUSIUM. In Chara one of the eight flat cells forming the wall of the globule.

SHIĔLD - SHĀPED, scutate or peltate.

SHOOT, a young growing branch or twig, or an unusually vigorous stem or branch, generally from the ground or near it. Also used by botanists as a general term for the stem or leaf-bearing part of a plant in distinction from the root.

SHRŬB, a woody plant which seldom exceeds twenty feet in height; especially one having several stems. Compare BUSH, UNDERSHRUB, and TREE.

SĬC'CŬS, dry.

SĬEVE-ÇĔLL, see SIEVE-TUBE.

SĬEVE-DĬSK, the perforated septum between the ends of adjoining sieve-tubes; sieve-plate.

SĬEVE-PLĀTE, see SIEVE-DISK.

SĬEVE-PŌREṢ, the perforations in a sieve-disk.

SĬEVE-TŪBE, a form of vessel characteristic of the phloëm portion of fibrovascular bundles. Sieve-tubes consist of elongated, thin-walled cells, united end to end in rows, with the transverse septa (sieve-disks) perforated in a sieve-like manner so that the protoplasm is continuous from one vessel to another; sieve-cell; clathrate-cell; bast-vessel.

SĬG'ĬLLĀTE, appearing as if marked with impressions of a seal, as the rootstock of Solomon's seal.

SĬG'MOID, curved in two directions, like the Greek sigma or letter S.

SĬL'ĬCLE, a short silique, as that of shepherd's-purse.

SĬLĬC'ŪLȦ (pl. **Sĭlĭc'ūlæ**), see SILICLE.

SĬL'ĬCŪLE, see SILICLE.

SĬLĬC'ŪLŌSE, having silicles, or resembling a silicle.

SĬL'ĬQUȦ (pl. **Sĭl'ĭquæ**), see SILIQUE.

SILIQUE [pro. **Sĭ-lēk'**], the fruit in Cruciferæ—a pod of two carpels, which separate in dehiscence from a frame-like placenta called a replum; siliqua.

SĬL'ĬQUŌSE, resembling or bearing siliques.

SĬLK, the style of Indian corn.

SĬLK'Y̆, see SERICEOUS.

SĬL'VȦ, see SYLVA.

SĬL'VĔR GRĀIN, plates of medullary rays which appear as glimmering spots upon the surface of wood when split radially. See MEDULLARY RAY.

SĬL'VĔRY̆, white, tinged with bluish gray, and having a metallic lustre.

SĬM'PLE, without subdivisions; entire; unbranched. Compare COMPOUND.

SĬM'PLE FRỤIT, one derived from a single flower containing a single pistil. Compare COLLECTIVE FRUIT and COMPOUND FRUIT.

SĬM'PLE LĔAF, any leaf which does not have distinct subordinate leaflets, though it may be lobed. Compare COMPOUND LEAF.

SĬM'PLE NŬTĀ'TION, see NUTATION. Compare CIRCUMNUTATION.

SĬM'PLE PĬS'TĬL, one consisting of a single carpel.

SĬM'PLE PĬT, one having no more than a slight enlargement at the centre where the pits of adjoining cells meet (simple pore of Gregory). Compare BORDERED PIT.

SĬM'PLE PŌRE (Gregory), see SIMPLE PIT.

SĬM'PLE PRĪ'MȦRY̆ ROOT, a tap-root.

SĬM'PLE SPŌ'RŌPHŌRE, a sporophore consisting of a single hypha or hyphal branch; filamentous sporophore. Compare COMPOUND SPOROPHORE.

SĬM'PLE STĔM, an unbranched stem.

SĬMŬLTĀ'NĔOŬS WHORL, one whose members originate simultaneously. (Sachs.) Compare SUCCESSIVE WHORL.

SĬN'GLE, said of a flower which has but one set or whorl of petals. Compare DOUBLE, SEMIDOUBLE, and FULL.

SĬN'ĬSTRŌRSE, turning or twining to the left; said of a twining stem which turns in the opposite direction to the sun or hands of a watch. See remarks under DEXTRORSE.

SĬN'ŪATE, strongly wavy on the margin, with large alternate convexities and concavities. Compare REPAND and UNDULATE.

SĬN'ŪOŬS, flexuose; curving back and forth.

SĪ'NŬS, a cavity or recess, either rounded or angular, in the margin of a leaf or other organ.

SĪ'PHŎN, a large tubular cell in Chara and various algæ.

SĬS'TĔR-ÇĔLL, a cell of the same origin as another: thus, if several cells arise simultaneously by free cell-formation within another, as in the development of pollen, they are called sister-cells with reference to each other, and the same term is applied to the relation existing between two or more cells which arise by the division of a single cell in ordinary growth. Compare MOTHER-CELL and DAUGHTER-CELL.

SKEIN, a condition of the chromatin of the nucleus in the first and last stages of karyokinesis when the nuclear filament is emerging from or passing into its reticulated condition in the resting nucleus; mother-coil; tangle.

SKĬN, any soft, thin covering, especially if easily removable; cuticle; epidermis. Compare RIND.

SLĂSHED, see LACINIATE.

SLĒEP, a condition assumed by certain plants, usually upon the withdrawal of light, as at night, in which the flowers temporarily close, and the leaves or leaflets droop or fold together.

SMALL FRUITS, a horticultural term for certain low-growing, perennial, fruit-bearing plants and their product, including the strawberry, raspberry, blackberry, gooseberry, currant, huckleberry, and cranberry. The term includes grapes, but excludes cherries.

SMOOTH, see GLABROUS.

SŌ'BŌLE, an old and useless term, usually meaning stolon, but sometimes and more properly a sucker or slender shoot from the ground or base of the stem; soboles.

SŎB'ŌLĒŞ (n., sing.), see SOBOLE.

SŌ'ÇIAL, growing naturally together, so as to occupy a considerable extent of ground with individuals of the same species; gregarious. Compare CESPITOSE.

SŌFT BÅST, the sieve-tubes, together with any other unlignified portion of the phloëm.

SŌ'LĒÆFÔRM, see CALCEOLATE.

SŎL'ĬD, without either cavities or separable articulations; continuous.

SŎL'ĬD BŬLB, see CORM.

SŎL'ĬTÁRY, single, not closely associated with other objects of the same kind, as flowers which grow one upon a stem, or plants which do not grow in groups or masses. Compare SOCIAL.

SŎLŪTE', free; not adhering: opposed to Adnate, as a *solute* stipule.

SOLU'TION, the isolation or separation of whorls (apostasis), or of organs or parts which are usually close together or adherent. Compare DIALYSIS.

SŌ'MÅ (pl. **Sō'måtå**), a small body or granule of any kind. Compare MICROSOMA.

SŌMĂT'ĬC CELLS, cells forming a part of the body of the individual, not modified for any particular purpose; especially the growth-cells of an organism, in distinction from reproductive cells of any kind.

SŌMÅTŎT'RŌPĬSM, an influence which the substratum exerts on the direction of growth of certain plants and organs. Thus the hypocotyl of the mistletoe is directed toward the branch (positive somatotropism) and the stem at first away from it (negative somatotropism) upon whatever side of the branch the seed is situated.

SŎR'DĬD, of a dirty white or muddy color.

SŌRĔ'DĬÅ, pl., see SOREDIUM.

SŌRĔ'DĬAL BRANCH, a branch produced by the development of a soredium into a new thallus while still on the mother-thallus.

SŌRĔ'DĬĀTE, see SOREDIFEROUS.

SŏRĒDĬF'ĒROŬS, bearing soredia; sorediate.

SŌRĒ'DĬŬM (pl. **Sōrē'diå**), a single gonidial cell or group of gonidia in lichens, surrounded by a weft of hyphæ, which is thrust out of the thallus and grows directly into a new lichen; brood-bud.

SŌ'RĪ, pl., see SORUS.

SŌRĬF'ĒROŬS, bearing sori.

SŌ'RŌSE, (1) heaped or clustered together; (2) bearing a sorosis, or in the form of a sorosis.

SŌRŌ'SĬS (pl. **Sōrō'sēs**), a collective fruit, consisting of a consolidated fleshy spike, as the mulberry and pineapple.

SŎR'RŌWFUL FLOW'ĒRS, those which exhale their odors only at certain hours of the day, as *Pelargonium triste*. (Rare.)

SŌ'RŬS (pl. **Sō'rī**), a heap or collection of spores or other objects, as a "fruit-dot" or group of sporangia in ferns.

SPÅDĬ'ÇEOŬS, resembling or bearing a spadix.

SPĀ'DĬCŌSE, resembling a spadix.

SPĀ'DĬX, a spike, usually fleshy, enclosed within a spathe.

SPĂN, the space between the tip of the thumb and that of the little finger, separated as widely as possible: nine inches.

SPÄRSE, few and scattered.

SPĀ'THÅ, see SPATHE.

SPÅTHÅ'ÇEOŬS, bearing or resembling a spathe; spathose.

SPĀ'THAL, furnished with a spathe. (Rare).

SPĀTHE, a large bract, usually colored, enclosing or subtending an inflorescence, which is generally a spike (spadix).

SPÅTHĔL'LÅ (pl. **Spåthĕl'læ**) (rare), (1) a glume in grasses; (2) a spathilla.

SPÅTHĬL'LA (pl. **Spåthĭl'læ**), a diminutive or secondary spathe, as in palms.

SPĂTH'ŌSE, see SPATHACEOUS.

SPĂTH'ŪLĀTE, see SPATULATE.

SPĂT'ŪLĀTE, flat, with a linear base and gradually rounded summit, like a druggist's spatula; spatulate.

SPAWN, the mycelium of mushrooms; occasionally applied to that of some other fungi

SPĒ'ÇIES, the lowest well-defined natural group of plants, the individuals of which differ but slightly among themselves and are sharply distinguished from those of other groups. They reproduce themselves from seed for successive generations with little or no variation. Compare GENUS and VARIETY.

SPĒ'ÇIES-HȲ'BRĬD, see HYBRID.

SPĒÇĬF'ĬC, relating to species.

SPĒÇĬF'ĬC CHĂR'ĂCTĒRS, the features which distinguish the plants of one species from those of all others.

SPĒÇĬF'ĬC NĀME, the name of the species, forming a part of the botanical name; trivial name.

SPĔÇ'ĬMEN, a plant or portion of a plant prepared and preserved for study.

SPĒR'MȦGŌNE, see SPERMOGONIUM.

SPĒRMȦGŌ'NĬŬM, see SPERMOGONIUM.

SPĒR'MȦPHŌRE, see SPERMOPHORE.

SPĒRMȦTĬF'ĒROŬS, bearing spermatia.

SPĒRMĀ'TĬŬM (pl. **Spērmā'tĭȧ**), a non-motile male gamete (spermatozoid) in the red algæ. Also applied to certain minute non-germinating spore-like bodies in fungi, some of which have been regarded as functionless male gametes. Many of these are now known to be *stylospores* capable of germination.

SPĒR'MȦTŌÇȲST, the mother-cell of a spermatozoid.

SPĒRMȦTŌZŌ'ĬD, see ANTHEROZOID and SPERMATIUM.

SPĒRMȦTŌZŌ'ŎN (pl. **Spermȧtōzō'ȧ**), see ANTHEROZOID.

SPĒRM-ÇĔLL, a male reproductive cell (gamete). Compare GERM-CELL.

SPĒRM'ĬC, pertaining to seed.

SPĒRMĬD'ĬŬM (pl. **Spĕrmĭd'ĭȧ**), an old term for achenium.

SPĒRM-NŪ'CLĒŬS, the nucleus of a male gamete (male pronucleus) which coalesces with the nucleus of a female gamete (female pronucleus) to form a germ-nucleus; spermo-nucleus. In the pollen-tube it is called the generative nucleus, which see.

SPĒR'MŌDĒRM, the covering of a seed, consisting of the several coats taken together; seed-coat.

SPĒRMŌGŌ'NĬŬM (pl. **Spērmōgō'nĭȧ**), a cell or receptacle in which spermatia are produced.

SPĒRMŌ-NŪ'CLĒŬS, see SPERM-NUCLEUS.

SPĒR'MŌPHŌRE, an old term, applied both to funiculus and placenta. See TROPHOSPERM.

SPĒR'MŌPHȲTE, a plant which produces true seeds instead of spores; a flowering plant.

SPĒRMŌTHĒ'CȦ, an old term for pericarp.

SPĒR'MOŬS, see SPERMIC.

SPHĂÇ'ĒLĀTE, dark and withered as though dying or dead; like a sphacelium.

SPHĂÇĒ'LĬȦ, a former genus of fungi, now known to be the first or conidia-bearing stage of Claviceps when it appears as the spur or ergot.

SPHȦÇĒ'LĬŬM, see SPHACELIA.

SPHǢRȦPH'ĬDĒS, sphere-crystals, especially if composed of needle-shaped parts.

SPHĒRE-CRȲS'TALS, spherical aggregations of crystals, either irregularly united or having a radiating structure. Compare

CYSTOLITH and SPHÆRAPHIDES.

SPHǼRŌ-CRȲS'TAL, see SPHERE-CRYSTAL.

SPĪ'CĀTE, arranged in a spike; spike-like; spicose.

SPĪÇĬF'ĒROŪS, bearing or producing spikes.

SPĪ'ÇĬFÔRM, spike-shaped.

SPĪCŌSE', bearing spikes or arranged in spikes; spicous.

SPĬC'ŪLA (pl. **Spic'ūlæ**), see SPICULE.

SPĬC'ŪLAR, resembling a spicule or bearing spicules.

SPĬC'ŪLĀTE, (1) covered with spicules; (2) bearing or composed of spikelets. (Rare.)

SPĬCŪLĀ'TION, a term employed by Nylander for an attenuated constriction of the hypha in the formation of spores, leaving the extremity as a spicule after the separation.

SPĬC'ŪLE, (1) a small pointed appendage somewhat soft or fleshy, like the sterigmata of fungi; (2) a needle-like crystal; (3) a spikelet (obs.); spicula.

SPĪKE, an inflorescence of sessile or nearly sessile flowers on a single elongated axis. See SPADIX, AMENT, SPIKELET, and COMPOUND SPIKE.

SPĪKE'LĔT, (1) a secondary spike; (2) in grasses a flower (or ultimate flower-cluster) ususally enclosed by one or more (generally two) empty glumes; locusta.

SPĬN'DLE, (1) a name sometimes given to the tassel of Indian corn; (2) see NUCLEAR SPINDLE.

SPĬN'DLE-FĪ'BREŞ, the achromatic filaments which form the nuclear spindle; spindle-threads; nuclear fibrils and cell-fibres of Strasburger; conjunctive threads of Fol (probably).

SPĬN'DLE-SHĀPED, see FUSIFORM.

SPĪNE, any sharp, rigid process of considerable size which is not a transformed branch; especially an organ, such as a leaf, stipule, tooth, etc., of whatever size, which is modified by becoming sharp and rigid for protection. In general, where no other distinction exists, as in the cactus, spines are considered merely as sharp processes intermediate in size and rigidity between thorns and prickles. See THORN and PRICKLE.

SPĪNĔS'ÇENT, somewhat spiny in structure, or bearing few spines. Compare SPINULESCENT.

SPĪNĬF'ĒROŬS, bearing a spine or spines; spinigerous.

SPĪN'ĬFÔRM, shaped like a spine or thorn.

SPĪNĬG'ĒROŬS, see SPINIFEROUS.

SPĪNŌSE', bearing spines; covered with spines; spinous; spiny.

SPĪ'NOŬS, spinose or spiny.

SPĬN'ŪLE, a very small spine.

SPĬNŪLĔS'ÇENT, slightly spiny in structure, or bearing small spines; spinulose. Compare SPINESCENT.

SPĬNŪLĬF'ĒROŬS, bearing very small spines.

SPĬN'ŪLŌSE, spinulescent; especially, bearing many small spines.

SPĪN'Ȳ, bearing spines; having the nature of a spine; terminating in a spine.

SPĪ'RAL DŬCT, see SPIRAL VESSEL.

SPĪ'RAL MÄRK'ĬNG, the spiral fibrous thickening characteristic of spiral ducts or vessels.

SPĪ'RAL VĔS'SĔL, a duct or vessel having fibrous thickenings upon the wall in the form of a coil or spiral. Formerly called Trachea.

SPĪRE, (1) a young leaf or shoot of grass; (2) the continuation of the trunk, in excurrent trees like pines, above the insertion of the lowest branches; (3) one turn of a coil or spiral.

SPĪ'RĬCLE, a minute spiral thread or filament.

SPĬTH'ÂMÂ (obs.), see SPAN.

SPLĂSHED (Hort.), having broken stripes of various sizes. Used mainly in describing the coloring of apples.

SPLĬT, the same as Parted. (Obs.)

SPŌN'ĠĬŌLE, a term formerly applied to the extreme apex of growing roots, which was erroneously supposed to be devoid of epidermis and specially fitted for the absorption of food from the soil. See ROOT-CAP.

SPŎNTĀ'NĒOŬS ĠĔNĔRĀ'TION, the development of living organisms from dead or inorganic matter; equivocal generation; heterogenesis; abiogenesis; autogenesis. The theory of spontaneous generation is not now held by scientists.

SPOON'FÔRM, having the inner surface of a leaf concave or dish-shaped, as the outer leaves of a cabbage-head. (E. S. Goff.)

SPŌRĂD'ĬC, scattered; occurring apart from others of the same kind.

SPŌRĂNĠE', see SPORANGIUM.

SPŌRĂNĠĬD'ĬŬM, (1) the columella in mosses; (2) a sporangium. (Rare.)

SPŌRĂN'ĠĬŌLE, an old term for Ascus. Also applied to a small sporangium produced in some genera of Mucorini in addition to the large form, the spores being similar in both; sporangiolum.

SPŌRĂNĠĪ'ŌLŬM, see SPORANGIOLE.

SPŌRĂN'ĠĬŌPHŌRE, the stalk, support, or receptacle of a sporangium, including such supports as the sporophyll in Equisetaceæ and the axis or columella from which the sporangia arise in the sori of certain ferns.

SPŌRĂNĠĬŌPH'ŌRŬM, see SPORANGIOPHORE.

SPŌRĂN'ĠĬŬM (pl. **Spōrăn'ġĭå**), any case or capsule immediately enclosing spores. Applied chiefly to certain kinds which have not received special names, such as the spore-cases of ferns and the Mucorini, and rarely used for Ascus, cluster-cup, and other named forms.

SPŌRE, one of the reproductive bodies of cryptogams which take the place of the seeds of flowering plants, but from which they differ in containing no embryo. Spores usually and properly consist of but one cell, which differs materially in character from the other cells of the plant. They may thus be distinguished from gemmæ, produced by comparatively few plants, which usually contain a number of cells only slightly differing from those of the plant producing them. In a restricted sense, proposed by Sachs, the term *spore* is applied to reproductive bodies which arise either directly or indirectly as the result of fertilization, in distinction from *gonidium*, which is applied to those produced asexually. Spores in this restricted sense are by De Bary termed *carpospores*. See ÆCIDIOSPORE, MACROSPORE, OÖSPORE, PROTOSPORE, TELEUTOSPORE, TETRASPORE, UREDOSPORE, ZOÖSPORE, ZYGOSPORE, etc.

SPŌRE-CĀSE, see SPORANGIUM.

SPŌRE-ÇĔLL, a spore, or a cell which gives rise to a spore.

SPŌRE-GROUP, see COMPOUND SPORE.

SPŌ'RĬD, see SPORIDIUM.

SPŌ'RĬDĔSM, see COMPOUND SPORE.

SPŌRĬD'ĬŬM (pl. **Spōrĭd'ĭà**), a spore produced upon a promycelium; germ-cell; secondary spore. Formerly applied to any spore, especially if very small, and still improperly used for Ascospore.

SPŌRĬF'ĔROŬS, spore-bearing.

SPŌ'RŌCÄRP, a group of spores resulting from fertilization, together with enclosing or attendant parts. It is thus the fructification developed from an archicarp or procarp in fungi and Rhodophyceæ; also the sporogonium in mosses. The term is also used for the body enclosing the sporangia in heterosporous Filicineæ. Compare SPOROGONIUM.

SPŌRŌCÄR'PĬŬM, see SPOROCARP.

SPŌRŌCLĀ'DĬŬM, a spore-bearing branch, as a stichidium in Florideæ.

SPŌ'RŌÇȲST, the mother-cell of a spore; sporocyte: occasionally also applied to certain sporangia in algæ.

SPŌ'RŌÇȲTE (Goebel), see SPOROCYST.

SPŌ'RŌDĔRM, the coat or covering of a spore, including exospore and endospore. Compare SPERMODERM.

SPŌRŌDŌ'CHĬŬM (pl. **Spōrōdō'chĭà**), a spore-bearing or spore-holding apparatus. Applied mainly to the sporiferous mass, including the spores, in the Tuberculariеæ.

SPŌ'RŌĠĔN, a plant which produces spores; cryptogam. Compare SPERMOPHYTE.

SPŌ'RŌGŌNE, see SPOROGONIUM.

SPŌRŌGŌ'NĬŬM (pl. **Spōrōgō'nĭà**), the whole product of the sexual act in the higher cryptogams; asexual generation; sporophore; sporophyte; oöphyte. In ferns it is the segment or stage in the life-cycle which we see as the ordinary fern-plant. Compare PROTHALLUS.

SPŌ'ROID, spore-like.

SPŌ'RŌPHŌRE, a stalk supporting one or more spores; any spore-bearing apparatus or receptacle, such as a gonidiophore, sporangiophore, or ascophore; particularly the second or asexual generation in plants having a distinct alternation of generations, such as ferns and mosses (sporogonium). Compare OÖPHORE. See SIMPLE SPOROPHORE, COMPOUND SPOROPHORE, and BASIDIUM.

SPŌ'RŌPHȲL, the same as Sporophyll or Sporophyllum and a spelling now in frequent use.

SPŌ'RŌPHȲLL, see SPOROPHYLLUM.

SPŌRŌPHȲL'LŬM (pl. **Spōrōphȳl'lȧ**), a spore-bearing leaf of any kind; sporophyl. Formerly applied to leaf-like lobes of the thallus in algæ bearing tetraspores.

SPŌ'RŌPHȲTE, see SPOROGONIUM.

SPŌRŌSTĔ'ĠIŬM (pl. **Spōrōstē'ġiȧ**), the oösporangium or fruit (nucule) of the Characeæ.

SPŌRT, a plant, or portion of a plant, which has suddenly assumed an appearance very different from that characteristic of the variety or species; a seed- or bud-variation, but said mainly of the latter. Its peculiarity is not usually capable of being perpetuated by seed. When a sport is propagated artificially it continues to be called a sport. Monstrosities or extreme malformations are not usually given this name. See VARIATION, SEED-VARIATION, and BUD-VARIATION.

SPŌRŬLĀ'TION, the production of spores.

SPŎR'ŪLE, a small spore. Often unsuitably applied to ascospores. Compare SPORIDIUM.

ṢPŎRŪLĬF'ĔROŬS, producing sporules.

SPŎRŪLĬĠ'ĔROŬS, see SPORULIFEROUS.

SPŌT'TĔD, covered with large dots: said of fruits.

SPRA̤WLS, a local or obsolete term for small branches or twigs.

SPRĔAD'ĬNG, said of branches which bend outward considerably, but at less than a right angle.

SPROUT, any quickly grown part, as a young root or stem from the seed, or a vigorous branch arising from the root or stem. See WATER-SPROUT.

SPROUT-ÇĔLL, in fungi, a cell produced by sprouting or pullulation.

SPROUT-CHĀIN, a row of united cells in fungi formed by budding (pullulation.)

SPROUT-ĠĔM'MȦ, in fungi, a gemma having the form of a septate confervoid filament the segments of which are capable of sprouting. (De Bary.)

SPROUT-ĠĔRMĬNĀ'TION, the manner of germination of a spore in which a small process with a narrow base protrudes at one or more points on the surface of the spore, then assumes an elongated cylindrical form, and is finally detached as a sprout-cell. (De Bary.)

SPROUT'ĬNG, see PULLULATING and GERMINATION.

SPŪMĔS'ÇENT, resembling froth or foam; spumose.

SPŪMŌSE', see SPUMESCENT.

SPÛR, (1) a short, stout branch, as those in the larch bearing a tuft of leaves or in the apple bearing the fruit; (2) a tubular appendage of a petal or sepal, usually containing nectar, as in larkspur (calcar); (3) the sclerotium of ergot.

SPŪ'RĬOŬS, see FALSE.

SPŪ'RĬOŬS BRȦNCH, see PSEUDORAMULUS.

SPŪ'RĬOŬS FRṲIT, a collective, aggregate, or accessory fruit.

SPŪ'RĬOŬS TĬS'SŪE, see FELTED TISSUE.

SPŪ'RĬOŬS WHŌRL, one which is formed by displacement and unequal growth of the axis. (Sachs.)

SPÛRRED, bearing a spur; calcarate.

SQUĀ'MÀ (pl. **Squā'mæ**), a scale of any kind, as one of the leaves of a bud.

SQUÀMĀ'ÇEOŬS, see SQUAMOSE.

SQUĀ'MÆFÔRM, see SQUAMIFORM.

SQUĀ'MĀTE, see SQUAMOSE.

SQUÀMĔL'LÀ (pl. **Squàmĕl'læ**), diminutive of Squama: a small scale, as the bracts upon the head in Compositæ; squamula.

SQUÀMĔL'LĀTE, diminutive of Squamate. See SQUAMULOSE.

SQUÀMĔLLĬF'ĔROŬS, bearing squamellæ.

SQUÀMĔL'LĬFÔRM, like a small scale; squamuliform.

SQUÀMĬF'ĔROUS, scale-bearing; squamigerous; squamose.

SQUĀMĬFLŌ'ROŬS, having flowers which resemble scales or are borne upon scales.

SQUĀ'MĬFÔRM, scale-shaped; squamoid.

SQUÀMĬG'ĔROŬS, see SQUAMIFEROUS.

SQUĀ'MŌID, see SQUAMIFORM.

SQUÀMŌSE', scale-like; covered with scales; consisting of scales; squamous; squamate; squamaceous. See SQUAMIFORM and SQUAMIFEROUS.

SQUĀ'MOŬS, pertaining to scales, or squamose.

SQUĀ'MOŬS BŬLB, see SCALY BULB.

SQUĂM'ŪLÀ, see SQUAMELLA. Formerly applied to the lodicule of grasses.

SQUĂM'ŪLĀTE, see SQUAMULOSE.

SQUĂM'ŪLĬFÔRM, see SQUAMELLIFORM.

SQUĂM'ŪLŌSE, diminutive of Squamose; having or resembling small scales; squamellate; squamulate.

SQUARRŌSE', having lateral organs, as leaves, extending at about right angles to the axis.

SQUĂR'RULŌSE, somewhat squatrose.

STĂG'-HĔADĔD, a term applied in forestry to a tree which is dying at the top.

STĀINED, faintly colored—applied to fruits.

STALKED GLÀND, see GLANDULAR HAIR.

STALK'LĔT, a secondary or very small stalk. See SECONDARY PETIOLE and SECONDARY PEDUNCLE.

STĀ'MĔN, the pollen-bearing organ of a flower, usually consisting of an anther, or part immediately enclosing the pollen, and a filament or stalk.

STĂM'ĬNAL (obs.), see STAMINATE.

STĂM'ĬNĀTE, pertaining to stamens; containing stamens but no pistils—said of a plant or flower. See STAMINEAL.

STÀMĬN'ĔAL, formed of stamens; attached to stamens; having marked reference to stamens; stamineous; staminate.

STÀMĬN'ĔAL CŎL'ŬMN, see ANDROPHORE.

STÀMĬN'ĔOŬS, see STAMINEAL.

STĂMĬNĬD'ĬŬM (pl. **Stămĭnĭd'ĭà**) (obs.), see ANTHERIDIUM.

STĂMĬNĬF'ĔROŬS, bearing stamens; staminigerous. Said especially of a plant or flower bearing stamens but no pistils.

STĂMĬNĬG'ĔROŬS, see STAMINIFEROUS.

STĂM'ĬNŌDE, see STAMINODIUM.

STĂMĬNŌ'DĬŬM (pl. **Stămĭnō'dĭà**), an altered, abortive, and sterile stamen, or body occupying the place of a stamen; staminode; parastemon. See LEPAL.

STĂM'ĬNŌDY̆, the conversion of other organs of a flower into stamens.

STĂM'ĬNŌSE, having the stamens unusually large or numerous. (Rare.)

STĂND'ARD, see VEXILLUM In horticulture (1) a tree or shrub which stands alone without being attached to any wall or support, as distinguished from an *espalier* or *cordon;* (2) a shrub, as a rose, grafted on an upright stem or trained to a single stem in tree form; (3) in the United States, a fruit-tree grafted upon a free-growing as opposed to one upon a dwarf stock.

STÄRCH-BUĬLD'ĔR, a plastid, generally a leucoplast, in which a starch-grain originates. See CHROMATOPHORE.

STÄRCH-GRĀIN, the grain of starch as it exists in the living plant. Ĭt is of definite shape, varying according to species, but always rounded, and formed of successive layers (or apparently so) having a common, and usually eccentric, centre.

STÄRCH-LĀY'ĔR, a term sometimes applied to the bundle-sheath, which often serves especially for the storage of starch.

STÄRCH-STÄR, a form of bulbil in *Chara stelligera*, consisting of an underground node.

STÄR'RY̆, see STELLATE.

STÄRVED, see DEPAUPERATE.

STĀ'SĬMÔRPHY̆, a changed condition of form or size arising from arrested development.

STĀ'TION, a particular limited locality, without regard to character or exposure, in which a species is found. Compare REGION, HABITATION and HABITAT.

STĂT'ŌSPŌRE, see RESTING-SPORE.

STĔGŌCĂR'POŬS, applied to capsules in mosses which have a distinct lid or operculum.

STĔL'LĀTE, star-shaped; radiating.

STĔL'LĀTE HÂIR, a hair having several branches arranged in the form of a star, as the hairs of Malpighiaceæ.

STĔL'LŪLAR, see STELLULATE.

STĔL'LŪLĀTE, diminutive of Stellate; resembling a little star; stellular.

STĔM, see TRUNK, PEDUNCLE, PEDICEL and PETIOLE.

STĔM-CLÅSP'ĬNG, see AMPLEXICAUL.

STĔM-LĔAF, a leaf growing from the stem above the ground. Compare RADICAL LEAF.

STĔM'LĔSS, see ACAULESCENT.

STĔM'LĔT, a little stem; the first internode above the cotyledons.

STĔM, SŬBTĔRRĀ'NĒAN, see SUBTERRANEAN STEM.

STĔM-TĔN'DRĬL, a tendril which is a transformed stem, as in the grape. Compare LEAF-TENDRIL.

STĔNŌPĔT'ALOŬS, having narrow petals.

STĔNŎPH'Y̆LLOŬS, narrow-leaved. Compare LATIFOLIOUS.

STĔNŌ'SĬS, (1) cell-formation in which there is a direct constriction of the walls of the original cell, as in budding. Compare CYTODIERESIS. (2) The pathological narrowing of a passage.

STĚ′RĚĬD, a cell of stereome, which see.

STĚ′RĚŌME, strengthening tissue: a term applied by Schwendener to that part of a fibrovascular bundle to which it chiefly owes its strength; also to the cortex on physiological grounds, it often serving mainly for support. Compare MESTOME.

STĚ′RĚŌPLĂSM, the solid portion of protoplasm. (Nägeli.) Compare HYGROPLASM.

STĚRĚŌPLĂS′MÀ, see STEREOPLASM.

STĚRĬG′MÀ (pl. Stērĭg′màtà), (1) the downward prolongation of a decurrent leaf upon the stem; (2) a very narrow support to a spore, as the pointed protuberances upon basidia, or the filaments bearing spermatia and stylospores.

STĚR′ĬLE, (1) unfruitful; barren —as a plant which fails to blossom or fruit, a pericarp which contains no seeds, or a stamen which produces no perfect pollen: (2) devoid of living organisms—used mainly in bacteriology.

STĚR′ĬLE FLOW′ĚR, one containing no perfect pistils. It may or may not contain stamens.

STĚR′ĬLĬZE, to render free from all living bodies, as fluids in which bacteria are to be cultivated.

STĚR′NŌTRĬBE, a term applied to zygomorphic flowers which have the stamens so placed that a visiting insect will receive the pollen upon its under surface, as in most Papilionaceæ. Compare NOTOTRIBE and PLEUROTRIBE.

STĚR′ŎM, see STEREOME.

STĬCHĬD′ĬŬM (pl. Stĭchĭd′ĭà), a branch of the thallus in red algæ (Florideæ) containing tetraspores.

STĬG′MÀ, the part of the pistil which receives the pollen. It is usually the apex of the style, variously expanded, which is destitute of epidermis and secretes the "stigmatic fluid," which see.

STĬG′MÀ-DĬSK, a disk forming the seat of a stigma, as in Asclepias.

STĬGMĂT′ĬC CĚLLS OF THE ĂRCHĒGŌ′NĬŬM, see LID-CELLS.

STĬGMĂT′ĬC CHĀM′BĚR, the part of the rostellum in orchids in which the viscid disk or retinaculum is developed. (Functionless as a stigma.)

STĬGMĂT′ĬC FLŪ′ĬD, a viscid fluid secreted by the stigma at maturity which serves to hold the pollen, and in which the pollen-grains germinate as the first stage of fertilization.

STĪLŌGŌNĬD′ĬŬM, see STYLOSPORE.

STĬM′ŪLŌSE, covered with stinging hairs.

STĬM′ŪLŬS (pl. Stĭm′ūlī), see STINGING-HAIR.

STĬNG, see STINGING-HAIR.

STĬNG′ĬNG-HÂIR, a glandular hair which secretes an acrid fluid; sting; stimulus.

STĪPE, a term applied to various kinds of stalks or stems, including the petiole of ferns, the stalk of a mushroom, the stem of a pappus above the seed, the stalk of an ovary raising it above the receptacle, etc. Various forms of stipes in flowering plants are distinguished under the terms *thecaphore, gynophore, antho-*

phore, gynobase, gonophore, carpophore.

STĪ'PĔL, the stipule of a leaflet.

STĬPĔL'LĀTE, furnished with stipels.

STĪ'PĒS, an uncommon spelling of Stipe.

STĪ'PĬFÔRM, resembling a stipe; stipitiform.

STĬP'ĬTĀTE, having a stipe; supported on a stipe.

STĬP'ĬTĬFÔRM, see STIPIFORM.

STĬP'ŪLĀ (pl. **Stĭp'ūlæ**), see STIPULE.

STĬPŪLĀ'ÇEOŬS, see STIPULAR.

STĬP'ŪLAR, having stipules; formed of stipules; situated near or upon stipules; pertaining to stipules; stipulaceous.

STĬP'ŪLĀRY̆, formed of stipules.

STĬP'ŪLĀTE, having stipules; stipulated; stipuled.

STĬP'ŪLĀTĔD, see STIPULATE.

STĬPŪLĀ'TION, the situation or arrangement of the stipules.

STĬP'ŪLE, an organ situated in pairs upon the stem, one on either side of the petiole at its base and often adherent to it. They are usually expanded, scale-like or leaf-like, but sometimes exist in the form of glands, prickles, tendrils, or other organs, and are often wanting altogether. They frequently serve for protection to the bud or growing point, and fall away as the bud expands and the leaves develop.

STĬP'ŪLED, see STIPULATE.

STĬPŪLĬF'ĔROŬS, bearing stipules.

STĬP'ŪLŌSE, having very large stipules. (Rare.)

STŎCK, (1) the persistent base of an herbaceous perennial; (2) any plant or part considered with reference to another part which it supports; (3) in horticulture a tree or other plant which receives a bud or graft; (4) the original or originals (parentage) of a species or variety; (5) a seedsman's term for a quantity of seed grown in a particular locality or from a particular source. It does not imply any peculiarity of character. Compare STRAIN.

STŌLE, see STOLON.

STŌ'LŌN, a slender branch, above or below the surface of the ground, which takes root or bears a bulb at the extremity where it forms one or more new plants; stole; sobole. Compare RUNNER, RHIZOME, and OFFSET.

STŎLŌNĬF'ĔROŬS, bearing stolons.

STŌ'MĀ (pl. **Stŏm'ātā**), an opening in the epidermis for the admission of air and liberation of moisture; breathing-pore. See GUARD-CELLS.

STŌ'MĀTE, see STOMA.

STŌMĂT'ĬC, pertaining to a stoma.

STŌMĂT'ĬC ÇĔLLS, see GUARD-CELLS.

STŌMĂT'ĬC CHĀM'BĔR, the intercellular space beneath a stoma; substomatic chamber; air-chamber; respiratory chamber; respiratory cavity. See CISTOME. Compare ANTECHAMBER and VESTIBULE.

STŎMĀTĬF'ĔROŬS, bearing stomata; stomatose.

STŎM'ĀTŌSE, see STOMATIFEROUS.

STŌNE, the bony endocarp of a drupe, containing the seed; pit.

STŌNE-FRUIT, see DRUPE.

STŌŌL, a plant upon which offshoots of any kind are produced, usually for propagation, as in the raspberry and strawberry; especially, a plant of wheat, oats, rye, or barley which has several stems from one root.

STŌŌL, v. i., to tiller, as grain; to throw out suckers.

STRĂG'GLĬNG, see DIVARICATE.

STRĀIN, (1) the influence of a particular ancestor existing in an individual or variety; (2) a seedsman's term for a stock of seed or subordinate variety which has been bred in a special manner and acquired a slight individual character, as "an excellent *strain* of Fottler's cabbage."

STRĂMĬN'ĔOŬS, like straw; especially, straw-colored; pale yellow.

STRĂND, a fibrovascular bundle; a branch of mycelium, or similar vegetable filament; specifically, a branch of fibrous mycelium, which see.

STRĂN'GŬLĀTĔD, applied to a root or stem which is contracted at intervals, as though growth had been restricted by cords or bandages wrapped around it. Compare MONILIFORM.

STRĂP, the ligule of a ray-floret in Compositæ.

STRĂP-SHĀPED, broad, flat, and linear, like a belt; ligulate.

STRĂTĬFĬCĀ'TION, the thickening of a cell-wall by the deposition of successive layers of formed material; also the arrangement of the layers so deposited.

STRĂT'ĬFĪED THĂL'LŬS, a thallus in lichens in which the gonidia or algal cells are disposed in one or more layers.

STRĀ'TŌSE, in distinct layers or strata.

STRAW, the stems of various edible grains, as wheat and oats, especially after being threshed. Extended also to the haulm of peas, beans, etc.

STRĔAK (Hort.), a long narrow stripe.

STRĔAM'ĬNG OF PRŌ'TŌPLĂSM, see ROTATION.

STRĪ'Ă (pl. Strī'æ), a straight and very narrow mark or groove.

STRĪ'ĀTE, marked with parallel striæ. Compare SULCATE, RIVULOSE, ACICULATED, and LINEATE.

STRĬCT, straight or upright; having branches close together and erect: opposed to Lax. Differs from Fastigiate in being applied chiefly to herbs and shrubs instead of trees and in not necessarily having a pointed top.

STRĪ'GĂ (pl. Strī'gæ), a sharp-pointed, rigid, appressed bristle or hair-like scale.

STRĬG'ĬLŌSE, minutely strigose.

STRĬGŌSE', covered with strigæ. Formerly regarded as synonymous with Hispid.

STRĪKE (Hort.), to take root, as a cutting.

STRĬNG, any fibre; strand.

STRĪ'ŌLĀTE, diminutive of Striate, having very fine striæ.

STRĪPED, having longitudinal lines of color, broader than Streaked and narrower than Banded; having longitudinal colored marks of any character.

STRŎBĬLĀ'ÇEOŬS, cone-bearing; cone-like; pertaining to cones.

STRŎB'ĬLE, see CONE.

STRŌBĬLĬF'ĔROUS, coniferous.

STRŌBĬL'ĬFŌRM, cone-shaped.

STRŌBĬ'LŬS, see CONE.

STRŌ'MÅ (pl. **Strō'måtå**), a mass in which another object is embedded, as a protein granule which contains a crystal or starch-grain. Especially. a compact mass of mycelium upon or within which spores or sporangia are borne—applied mainly to tissue in which perithecia are embedded. Compare HYMENIUM.

STRŌ'MÅTŌID, having the nature or appearance of a stroma.

STRŌ'MÅTOŬS, bearing or producing stromata.

STRŌMBŪLĬF'ĔROŬS, bearing spirally twisted pods (strombi).

STRŌMBŬ'LĬFŌRM, twisted in a spiral, like the pods of alfalfa; cochleate.

STRŎM'BŬS (pl. **Strŏm'bī**), a spirally coiled legume, like that of alfalfa, *Medicago sativa;* cochlea.

STRŌ'PHĒ, an old term for leaf-spiral.

STRŌ'PHĬŌLĀTE, furnished with a strophiole.

STRŌ'PHĬŌLE, see CARUNCLE.

STRŬC'TŪRAL BŎT'ÅNY̆, the study of the form and arrangement of the parts of plants, including morphology and histology; vegetable anatomy; phytotomy.

STRṲ'MÅ, a one-sided swelling or protuberance of any kind, as the one-sided enlargement (apophysis) at the base of the capsule in some mosses, or the pulvinus of a leaf or leaflet. The term is now seldom used except in mosses.

STRṲ'MĬFŌRM, like a struma in appearance.

STRṲMŌSE', swollen upon one side like a goitre; bearing a struma.

STRṲ'MŪLŌSE, slightly strumose.

STŬFFED, the same as Solid, or the old term Farctate, but seldom applied except to stems having a fibrous or spongy centre. Used mainly in mushrooms.

STŬMP'-ROOTĔD, applied in vegetable gardening to the roots of certain varieties of the radish, carrot, etc., which, instead of tapering gradually from the fleshy edible portion to the slender tap-root, have the transition abrupt. The term is not applied to the roots of turnip beets, flat turnips, etc., but only to such as are more elongated.

STŪ'PÅ, a tuft of matted hairs resembling tow. (Obs.)

STŪ'PĒOŬS, see STUPOSE.

STŪPŌSE', resembling tow; covered with matted hairs resembling tow; stupeous.

STȲLE, the slender part of a pistil supporting the stigma.

STȲLĬF'ĔROŬS, style-bearing.

STȲ'LĬFÔRM, style-shaped.

STȲLŌGŌNĬD'ĬŬM, see STYLOSPORE.

STȲ'LŌPŌD, see STYLOPODIUM.

STȲLŌPŌ'DĬŬM, the fleshy disk crowning the ovary in most Umbelliferæ, formed by the expanded bases of the two styles.

STȲLŌSE', having the styles remarkably long, numerous, or persistent. (Rare.)

STȲ'LŌSPŌRE, a stalked spore, sometimes septate, produced in a pycnidium; stylogonidium; pycnogonidium; pycnospore; pycnidiospore; clinospore.

STYLŌSTĔ'ĠĬŬM (pl. **Stylōstē'ġĭă**), the peculiar orbicular corona, called scutum or shield, which terminates the style in Stapelia and similar asclepiads.

STY̆LŌSTĔ'MŌN, an epigynous stamen. (Rare.)

SUĀVĔ'ŌLENT, sweet-scented.

SUB-, a Latin prefix meaning under, beneath, subordinate, or partially.

SŬBȦCŬTE', moderately acute.

SŬBĀĔ'RĬAL, situated just above the ground or substratum, but covered with fallen leaves, etc., as many rhizomes.

SŬBȦPĬC'ŪLĀTE, having a small or poorly defined apiculus or point.

SŬBÄRBŌRĔS'ÇENT, somewhat tree-like.

SŬBĂX'ĬLLĀRY̆, situated beneath the axil.

SŬBCA̤ULĔS'ÇENT, with the leafy stem so short as to be hardly noticeable, but rather more developed than in Acaulescent.

SŬBÇĔS'PĬTŌSE, somewhat inclined to grow in bunches. See CESPITOSE.

SŬB'CLȦSS, a group of orders or cohorts next subordinate to a class.

SŬBCŎNCĂT'ĒNĂTE, growing somewhat in chains—in short or imperfect rows or chains.

SŬBCŌN'ĬCAL, slightly conical.

SŬBCŌNTĬN'ŪOŬS, rarely or imperfectly septate; seldom or slightly constricted.

SŬBCŌR'DĀTE, slightly cordate.

SŬBCRĔ'NĀTE, slightly or obscurely crenate.

SŬBCŬL'TRĀTE, slightly cultrate.

SŬBDĔN'TĀTE, imperfectly or obscurely dentate. Compare DENTICULATE.

SŬBDĔNTĬC'ŪLĀTE, obscurely denticulate: having small, imperfect marginal teeth.

SŬBĔFFŪSE', having slightly spreading branches or filaments.

SŬBĔNTĬRE', only slightly notched or toothed.

SŬBĔPĬDĔR'MAL TĬS'SŪE, see HYPODERMA.

SŪ'BĔR, see CORK.

SŬBĔRĔCT', nearly erect.

SŬBĒ'RĔOŬS, see SUBEROSE.

SŪ'BĔRĬN, cork-substance—nearly or quite the same as Cutin. See CORK.

SŪBĔRĬFĬCĀ'TION, see SUBERIZATION.

SŪBĔRĬZĀ'TION, conversion into cork; suberisation; suberification.

SŬBĔRŌSE', having somewhat the appearance of being gnawed; slightly erose. Compare EROSE.

SŪ'BĔRŌSE, of the nature of cork; suberous.

SŪ'BĔROŬS, see SUBEROSE.

SŬBFĂM'ĬLY̆, a group of genera subordinate to a family.

SŬBFLĔX'ŪŌSE, slightly wavy.

SŬBĠĔNĬC'ŪLĀTE, slightly bent.

SŬBĠĒ'NŬS, a species, or group of species, having nearly the rank of genus.

SŬBGLŌBŌSE', nearly globose.

SŬBHY̆MĒ'NĬŬM, a layer of tissue next below the hymenium; hypothecium.

SŬBĬC'ŪLŬM, a term applied by M. C. Cooke to the abnormally thickened portion of a leaf or stem bearing the cluster-cups in Æcidium.

SŬBKĬNG′DŎM, the highest subdivision of a kingdom, as the phenogams and cryptogams among plants; series. Some recent botanists divide the vegetable kingdom into more than two subkingdoms.

SŬBLĔNTĬC′ŪLAR, somewhat lens-shaped.

SŬBMÄR′ĠĬNAL, situated near the margin.

SŬBMĒRĠED′, see SUBMERSED.

SŬBMĒRSED′, growing under water; submerged; demersed.

SŬBNĂS′ÇENT, originating or growing beneath any object.

SŬBNŪDE′, nearly destitute of hairs, leaves, or other covering.

SŬBŎBTŪSE′, slightly obtuse.

SŬBÔRBĬC′ŪLAR, nearly circular; subrotund. Compare SUBGLOBOSE.

SŬBÔR′DĒR, a group of genera within an order. Compare TRIBE.

SŬBŌ′VATE, somewhat ovate.

SŬBPĒDŬN′CŪLĀTE, having a very short peduncle; subsessile.

SŬBPĔT′ĬŌLAR, situated beneath the petiole, i.e., under (within) the base of the petiole, as the leaf-bud in sycamore. Compare SUBPETIOLATE.

SŬBPĔT′ĬŌLĀTE, having a very short petiole; subsessile. Compare SUBPETIOLAR.

SŬBRĀ′MŌSE, having very few branches; subsimple. Compare RAMOSE.

SŬBRĬĠ′ĬD, slightly rigid.

SŬBRŌTŬND′, see SUBORBICULAR.

SŬBSĔR′RĀTE, sparsely or obscurely serrate. Compare SERRULATE.

SŬBSĔS′SĬLE, nearly sessile; subpetiolate or subpedunculate.

SŬBSHRŬB′BY̆, see FRUTESCENT.

SŬBSĬD′ĬARY̆ ÇĔLL, see ACCESSORY CELL.

SŬBSĬM′PLE, with very few subdivisions or branches.

SŬBSPĒ′ÇIĒŞ, a variety nearly equal in rank to a species; a marked variety.

SŬB′STĬTŪTE FĪ′BRE, a term applied by Sanio to a prosenchymatous cell larger in diameter than a libriform cell and without its attenuated ends. Not in general use.

SŬBSTŌMĂT′ĬC CHĀM′BĒR, see STOMATIC CHAMBER.

SŬBSTRĀ′TŌSE, in somewhat indistinct or irregular layers or strata.

SŬBTĔND′, to embrace in its axil. A flower is *subtended* by a bract when there is a bract situated beneath it. Nearly all buds are *subtended* by leaves.

SŬBTĒRĒTE′, somewhat terete.

SŬBTĔRRĀ′NĒAN, beneath the surface of the ground; hypogean.

SŬB′TRĪBE, a group of genera subordinate to a tribe.

SŪ′BŪLĀTE, narrow, and tapering to a sharp rigid point, as the leaves of juniper; awl-shaped.

SŪBŪLĬF′ĒROŬS, furnished with awl-shaped spines. (Obs.)

SŪ′BŪLĬFÔRM (obs.), see SUBULATE.

SŬBŬM′BĔLLĀTE, somewhat umbelliform, as the flower-cluster of the apple.

SŬBVȦRĪ′ĔTY̆, a subordinate variety; a variety within a variety.

SŬBVĔN′TRĬCŌSE, somewhat ventricose or inflated.

SŬBVÊRTĬÇ'ĬLLĀTE, slightly verticillate; in imperfect or irregular whorls.

SŬCÇĔDĀ'NĔŬM, a substitute.

SŬCÇĔS'SĬVE WHŌRL, a whorl the members of which do not all originate at the same time, but in succession, either in regular order or otherwise. Compare SIMULTANEOUS WHORL.

SŬCÇĬF'ĔROŬS, producing or conveying sap.

SŬCÇĪSE', terminating abruptly, as though cut sharply off. Compare TRUNCATE and PREMORSE.

SŬC'CŌSE, see SUCCULENT.

SŬC'CŪBOŬS, having the base or lower edge of each leaf overlapping the apex of the preceding leaf, as in Jungermannia. Compare INCUBOUS.

SŬC'CŪLENT, soft, and fleshy or juicy; pulpy.

SŬCK'ĔR, (1) a sprout or shoot, especially from the root or lower part of the stem (surculus); (2) sometimes applied to aërial roots or other holdfasts; (3) a haustorium.

SŬCK'ĔR, v. i., to put forth shoots from the lower part of the stem, as Indian corn.

SŬDORĬF'ĬC, causing perspiration.

SŬFFRUTĔS'ÇENT, slightly shrubby.

SŬF'FRUTĔX, a plant with a woody base and herbaceous stems or branches. Applied mainly to perennials. See UNDERSHRUB.

SŬFFRU'TĬCŌSE, either suffrutescent or like an undershrub.

SŬFFRUTIC'ŬLŌSE, slightly fruticulose, as some lichens.

SŬL'CĀTE, having one or more large, straight, longitudinal grooves or channels, as the stem of parsnip. Compare RIVOSE and STRIATE.

SŬL'CŬS (pl. Sŭl'çī), a longitudinal groove or furrow; sulcation. Compare STRIA.

SŬLPHŪ'RĒOŬS, sulphur-yellow —paler than Luteus. About the same as Flavus.

SŬM'MĔR-SPŌRE, any spore or gonidium which retains its vitality but a short time, and is intended for the propagation of the plant during the summer, as the uredospores of wheat-rust. Compare RESTING-SPORE.

SUPER- (or **SUPRA-**), a prefix from the Latin, meaning above in position or degree.

SŪPĔRĂX'ĬLLĀRY̆, situated above the axil, instead of within it, as the accessory buds sometimes seen in the butternut; supraaxillary; superfoliaceous; suprafoliaceous.

SŪPĔRCRĔS'ÇENCE, a parasite. (Rare.)

SŪPĔRCRĔS'ÇENT, growing above another part or body; especially, growing upon some other growing thing. (Rare.)

SŪPĔRDĒCŎMPOUND', see SUPRADECOMPOUND.

SŪPĔRFĔCŬNDĀ'TION, the conjugation of more than two gametes. Compare SUPERFETATION.

SŪPĔRFĒTĀ'TION, the fertilization of an ovary by two or more kinds of pollen, so that its seeds are not all alike. Compare SUPERFECUNDATION.

SŪPĔRFŌLĬĀ'ÇEOŬS, see SUPERAXILLARY.

SŪPĒ'RĬOR, said of a radicle when it points towards the

apex of the fruit (ascending); said of the ovary when the calyx is free from it; said of the calyx when adherent to the ovary and thus, with the other floral organs, situated upon its summit; applied to the upper side of a lateral flower, the side next the axis: the vexillum, for example, is the *superior* petal of a papilionaceous corolla.

SŬPĔRNĀ'TANT, floating on the surface.

SŪPĔRNŪ'MĔRĀRY̆ BUDS, see ACCESSORY BUDS.

SŬ'PĔRPĂRĂSĪTE, a parasite of a parasite; hyperparasite.

SŬ'PĔR-PLĂNT, a plant which grows upon another plant, either as an epiphyte or a parasite.

SŪPĔRPŌṢĬ'TION, situation above some other organ; or, when referring to parts of the flower, the non-alteration of the members of contiguous circles, the corresponding parts being opposite instead of alternate; anteposition. See DIRECT and INVERTED SUPERPOSITION.

SŪPĔRTŪBĔRĀ'TION, an abnormal production of secondary tubers directly upon those produced in the ordinary manner.

SŪPĔRVŌLŪTE', plaited and the plaits convolute, as the corolla of morning-glory.

SŪPĪNE', lying flat, with face upward; dorsicumbent—opposed to Prone. Compare RESUPINATE.

SŬPPŌRT'ĬNG PLĂNT, a plant upon or within which another grows, either as a parasite or an epiphyte. See HOST.

SŬPPRĔS'SION, the entire non-development of a part; complete abortion; obliteration.

SUPRA-, see SUPER-.

SŪPRĂĂX'ĬLLĀRY̆, see SUPERAXILLARY.

SŪPRĂDĒCŌM'POUND, three or more times compounded, or very much subdivided in any manner. A pinnate leaf, like that of walnut, is *compound;* a bipinnate leaf, like that of honey locust, is *decompound;* a tripinnate leaf, as in many acacias, is *supradecompound.*

SŪPRĂFŌLĬĀ'ÇEOŬS, see SUPERAXILLARY.

SŪPRĔME', situated at the highest point.

SÛRCŪLĬĠ'ĔROŬS, see SURCULOSE.

SÛR'CŪLŌSE, having suckers (surculi).

SÛR'CŪLŬS (pl. **Sûr'cūlī**), a sprout or shoot from the ground or base of the stem; sucker.

SŬRCŬR'RENT, having winged expansions from the base of the leaf prolonged up the stem. Compare DECURRENT.

SŬSPĔND'ED, said of seeds or ovules which are attached to the top or sides of the ovary and hang downward; pendulous; inverted. Compare ERECT and ASCENDING.

SŬSPĔN'SOR, (1) a single or multiple row of cells which is the first development of the fertilized oösphere in phenogams, and at the extremity of which the embryo is developed: it appears upon the embryo as a continuation of the radicle; proembryo; (2) the cell which supports the conjugating cell in Mucorini.

SŬ'TŪRE, a line of junction or dehiscence.

SWARM, a number of spores or unicellular individuals of common origin which remain near together without being united in any way.

SWARM-CĔLL, see ZOÖSPORE.

SWARM'ĬNG, moving about by means of cilia—said of zoöspores. Chiefly used when many are together. See SWIMMING.

SWARM-SPŌRE, see ZOÖSPORE.

SWĬM'MĬNG, moving unattached within a liquid in any definite manner. Compare NATANT. The swimming of zoöspores when massed, however, is termed Swarming.

SWŌRD'-SHĀPED, see ENSIFORM,

SY̆CHNŌCÄR'POŬS, see POLYCARPIC.

SȲCŌ'NĬŬM, see SYCONUS.

SȲCŌ'NŬS, the fruit (hypanthium) of the fig, consisting of an expanded fleshy receptacle enclosing the flowers.

SY̆L'VÀ (pl. **Sy̆l'væ**), the trees of a country or region, or a work describing them.

SY̆L'VAN, pertaining to woods.

SY̆LVĔS'TRĬNE, growing in woods.

SY̆MBĪŌ'SĬS, the coexistence in more or less mutual interdependence of two different organisms, as the fungus and alga which constitute a lichen; mutualism; mutual parasitism; commensalism; consortism. With some authors Commensalism implies an association less necessary or mutually helpful than Symbiosis.

SY̆MMĔT'RĬCAL, (1) having sepals, petals, and stamens of the same number, or multiples of one another; (2) divisible in one or more directions into halves, which are similar to, or the reflections of, each other. See MONOSYMMETRICAL and POLYSYMMETRICAL.

SY̆MPĔT'ALOŬS, see GAMOPETALOUS. Also formerly used for a partial union of the petals with monadelphous stamens, as in Malvaceæ.

SY̆MPHY̆ĂN'THÊROŬS, see SYNANTHEROUS.

SY̆MPHY̆CÄR'POŬS, having the fruits confluent, as the disks of the apothecia in certain lichens.

SY̆MPHY̆L'LOŬS, see GAMOPHYLLOUS.

SY̆MPHY̆ŌĠĒNĔT'ĬC, formed of parts which have become grown together or united.

SY̆MPHY̆ŌSTĔM'ŌNOŬS, monadelphous, synantherous, or with the stamens united in any other manner: symphystemonous.

SY̆M'PHY̆SĬS, a union of parts usually distinct; cohesion or adhesion.

SY̆MPHY̆STĔM'ŌNOŬS, see SYMPHYOSTEMONOUS.

SY̆M'PŌDE, see SYMPODIUM.

SY̆MPŌ'DĬAL DĪCHŎT'ŌMY̆, where one branch of each successive bifurcation continues to develop and the other remains subordinate. See SCORPIOID and BOSTRYCHOID DICHOTOMY.

SY̆MPŌ'DĬŬM, a stem which consists of a series of secondary stems or axes which have arisen as branches one from another, as in the tomato; pseudaxis; false axis.

SY̆NĂC'MY̆, having the stamens and pistils ripen at the same time, neither protandrous nor protogynous: synanthesis: opposed to Heteracmy.

SYNAN'GIUM, the peculiar boat-shaped sorus of certain ferns of the order Marattiaceæ.

SYNAN'THEROUS, having the stamens united by their anthers, as in Compositæ; symphyantherous; syngenesious.

SYNANTHE'SIS, see SYNACMY.

SYNAN'THOUS, (1) having the flowers expand at the same time as the leaves. Compare PROTERANTHOUS and HYSTERANTHOUS. (2) Exhibiting synanthy.

SYNAN'THY, the abnormal coalescence of two or more flowers.

SYN'CARP, see AGGREGATE FRUIT.

SYNCAR'PIUM, see AGGREGATE FRUIT.

SYNCAR'POUS, formed of distinct aggregated carpels.

SYNCOTYLED'ONOUS, having coherent cotyledons.

SYNCY'TIUM, a collection of nuclei without cell-walls.

SYNE'DRAL, growing upon the angle of a stem.

SYNE'MA, that part of the column in orchids which represents the united filaments of the stamens.

SYNER'GIDÆ (sing. **Syner'gida**), two cells (or nuclei) in the upper end of the embryo-sac, which together with the oösphere form the egg-apparatus.

SYN'ERGY, the simultaneous action of several organs.

SYNGENE'SIOUS, see SYNANTHEROUS.

SYNOC'REATE, said of stipules when united around the stem so as to form an ocrea or sheath.

SYNŒ'CIOUS, (1) having male and female flowers in the same head, as in some Compositæ; (2) having antheridia and archegonia in the same receptacle in mosses.

SYN'ONYM, a superseded plant-name.

SYNOPH'YTY, the cohesion of two or more embryos in a seed. (M. C. Cooke.)

SYNOP'SIS (pl. **Synop'ses**), a condensed description of a genus, species, or other group.

SYNPET'ALOUS, see GAMOPETALOUS.

SYNSEP'ALOUS, see GAMOSEPALOUS.

SYNTAG'MA (pl. **Syntag'mata**), a name applied by Pfeffer to all bodies composed of tagmata, which see.

SY'PHON, see SIPHON.

SYS'TEM, (1) an arrangement of natural objects according to some rule; (2) the sum of the parts of an organism which are of the same morphological nature or perform a similar function, as the fibrovascular or intercellular system.

SYS'TEM, ARTIFI'CIAL, see ARTIFICIAL SYSTEM.

SYSTEMAT'IC BOT'ANY, the part of botany which treats of the description, naming, and classification of plants. See VEGETABLE TAXONOMY and PHYTOGRAPHY.

SYS'TEM, NAT'URAL, see NATURAL SYSTEM.

SYS'TROPHE, the massing of the chlorophyll bodies of a cell under intense light. Compare APOSTROPHE and EPISTROPHE.

TABES'CENT, wasting or shrivelling.

TAB'ULAR, flattened horizontally.

TAG'MA (pl. **Tag'mata**), a name given by Pfeffer to any aggregate of molecules, including pleon, micella, and micellar aggregate. See these terms.

TĀIL, any long, flexible, terminal appendage, as the persistent style on the seed of Clematis.

TĀIL'-POINTĔD, tipped with a long, flexible acumination.

TĂN'GLE, see SKEIN.

TĂNK'ARD-SHĀPED, thickened, about twice as long as broad, gradually enlarged downward, then suddenly contracted or terminated, as the root of some varieties of the turnip and radish. Compare STUMP-ROOTED.

TĀ'PĔR-POINTĔD, see ACUMINATE.

TȦPĔ'TŬM, (1) a layer of cells, just outside the archesporium, lining the cavity of an anther or a sporangium. It usually becomes disorganized and absorbed before the liberation of the spores or pollen-grains; (2) a similar layer of cells surrounding the embryo-sac.

TĂP'-RŌOT, a main root which runs directly downward.

TÄRTĀ'RĔOŬS, having the surface rough and crumbly, as that of many lichens. (Obs.)

TĂS'SEL, the popular name for the staminate inflorescence or terminal compound spike of Indian corn. Sometimes called Spindle.

TAW'NȲ, see FULVOUS.

TĂXŌL'ŌĠȲ, see TAXONOMY.

TĂXŌN'ŌMȲ, see VEGETABLE TAXONOMY.

TĒAR, a drop of gum or resin as it has issued from the plant.

TĒAR'-SHĀPED, of the shape of an apple-seed—the same as pear-shaped except that the sides are not contracted.

TĒETH, see TOOTH.

TĔG'MĔN, the innermost seed-coat; endopleura. Called Secundine in the ovule.

TĔGMĔN'TŬM (pl. Tĕgmĕn'tȧ), an old term for Bud-scale.

TĔG'ŪMENT, see INTEGUMENT.

TĒ'LȦ CŎNTĔX'TȦ, see FELTED TISSUE.

TĔLEŪ'TŌSPŌRE, a thick-walled, usually compound, gonidium produced by the Uredineæ or rust-fungi late in the season and which serves to reproduce the fungus the next year; brand-spore; pseudospore.

TĔN'DRĬL, a slender appendage which serves for support by coiling around some other object. It may be morphologically a leaf, leaflet, stipule, or stem.

TĔN'TȦCLE, one of the sensitive glandular hairs on the leaf of Drosera.

TĔNŪĬFŌ'LĬOŬS, having thin, narrow leaves.

TĔP'AL, one of the parts of a perianth, either sepal or petal. (Rare.)

TĔRȦTŎL'ŌĠȲ, the study of abnormal structures; morphology as applied to monstrous growths. Not applied to malformations due to disease.

TĔR'ÇĬNE, a third coat to the ovule, counting from the outside—not a constant and definite structure.

TĔRĒTE', cylindrical, or somewhat tapering.

TĔRĠĔM'ĬNAL, see TERGEMINATE.

TĔRĠĔM'ĬNĀTE, having three pairs of leaflets or other organs attached, by secondary petioles or otherwise, to the apex of a common support. Compare

TERNATE, TERNATE-PINNATE, and TRIJUGATE.

TĔRĠĬF'ĔROŬS (obs.), see DORSIFEROUS.

TĔRĠĬSPĔR'MOŬS, bearing reproductive bodies upon the back, as the leaves of ferns. (Obs.)

TĔR'GŬM (obs.), see DORSUM.

TĔR'MĬNAL, attached to or pertaining to the extremity or apex.

TĔR'MĬNAL BŬD, a bud at the extremity of a branch or stem. Compare LATERAL BUD.

TĔR'NÀRY̆, consisting of three. Compare TERNATE.

TĔR'NĀTE, growing in threes, as the leaflets in clover.

TĔR'NĀTELY̆ TRĪFŌ'LĬŌLĀTE, having three leaflets attached at one point, as in clover. Compare TRIFOLIOLATE.

TĔR'NĀTE-PĬN'NĀTE, having three secondary petioles, each bearing pinnate leaflets, attached to the apex of a common petiole. If each secondary petiole bears but a single pair of leaflets the leaf is called Tergeminate.

TĔRPĬN'NĀTE, see TRIPINNATE.

TĔRRĔS'TRIAL, growing on land, not aquatic; growing on the ground, not on trees.

TĔS'SĔLLĀTĔD, having square spots, arranged like those on a chess-board.

TĔS'TÀ, the outer seed-coat, called primine, in the ovule; spermoderm.

TĔSTĀ'ÇEOŬS, brownish yellow, like unglazed earthen-ware.

TĔSTĬC'ŪLĀTE, oval and solid, like the tuberous roots of certain orchids.

TĔST'ŪLE (obs.), see FRUSTULE.

TETRA-, in Greek compounds, four.

TĔTRÀCĂM'ÀROŬS (obs.), about the same as Tetracoccous. See CAMARA.

TĔTRÀCÄR'PĔLLÄRY̆, of four carpels.

TĔTRÀCHĒ'NĬŬM, a fruit consisting of four achenium-like carpels, as in Labiatæ.

TĔTRÀCHŎT'ŌMOŬS, dividing at the end into four branches.

TĔTRÀCŎC'COŬS, of four cocci.

TĔTRÀÇY̆C'LĬC, composed of four whorls, as a flower having calyx, corolla, and stamens each of one whorl and a single (simple or compound) pistil.

TĔT'RĂD, a group of four pollen-grains.

TĔTRÀDY̆N'ÀMOŬS, having six stamens, four of which are longer than the other two. Compare DIDYNAMOUS.

TĔTRÀFŌ'LĬOŬS, having bijugate leaves, i e., having leaves with two pairs of leaflets.

TĔTRĂG'ŌNAL, prismatic and quadrangular; four-angled; tetragonous. Compare TETRAQUETROUS.

TĔTRÀGŌNĬD'ĬŬM (pl. **Tĕtràgō-nĭd'ĭà**), see TETRASPORE.

TĔTRĂG'ŌNOŬS, see TETRAGONAL.

TĔTRĂĠ'Y̆NOŬS, having four pistils or styles.

TĔTRĂM'ĔROŬS, having four parts or the parts in fours. Applied mainly to flowers which have four members in each set or whorl of organs. Compare DIMEROUS, TRIMEROUS, etc.

TĔTRĂN'DROŬS, having four stamens.

TĔTRȦPĔT'ALOŬS, having four petals.

TĔTRĂPH'ȲLLOŬS, four-leaved. Often used for Tetrasepalous.

TĔTRĂQ'UĔTROŬS, having four salient angles. Compare TETRAGONAL.

TĒ'TRÄRCH, a term applied to a fibrovascular cylinder which represents four fibrovascular bundles.

TĔTRȦSĔP'ALOŬS, of four sepals. See TETRAPHYLLOUS.

TĔTRȦSPĔR'MOŬS, four-seeded.

TĔTRȦSPŌRĂN'ĠĬŬM, a sporangium (unicellular) containing tetraspores.

TĔT'RȦSPŌRE, a gonidium or asexually produced spore in Florideæ. So called from being often produced four together in a mother-cell; tetragonidium.

TĔTRĂS'TĬCHOŬS, in four vertical rows upon a stem.

THĂLȦMĬFLŌ'RAL, having the stamens inserted upon the receptacle. Compare COROLLIFLORAL and CALYCIFLORAL.

THĂLȦMĬFLŌ'ROŬS, see THALAMIFLORAL.

THĂL'ȦMŬS (pl. Thăl'ȧmī), the receptacle of a flower.

THȦLĂS'SŌPHȲTE, a sea-alga. (Rare.)

THĂL'LŌĠĔN, see THALLOPHYTE.

THĂL'LOĪD, having the form or nature of a thallus. Compare FOLIOSE and FRONDOSE.

THĂLLOĪ'DAL, see THALLOID.

THĂL'LŌME, see THALLUS. Compare CAULOME.

THĂL'LŌPHȲTE, a plant whose vegetative body is a thallus, as a lichen, fungus, or alga; cellular cryptogam; thallogen. Compare CORMOPHYTE.

THĂL'LŬS (pl. Thăl'lī), a vegetative body without true leaf or stem, as that of most cryptogams; thallome.

THĂL'LŬS PLȦCŌ'DĒS (obs.), see FOLIACEOUS THALLUS.

THĂL'LŬS THĂMNŌ'DĒS (obs.), see FRUTICOSE THALLUS.

THĒ'CȦ (pl. Thē'çæ), a sporangium or anther-cell. (Rare.) Formerly in general use for ascus, and still used by some writers for the sporangium or capsule of mosses.

THĒ'CȦPHŌRE (obs.), see GYNOPHORE.

THĒ'CȦSPŌRE, see ASCOSPORE.

THĒÇĬF'ĔROŬS, bearing asci or other thecæ; thecigerous. (Rare.)

THĒÇĬĠ'ĔROŬS, see THECIFEROUS.

THĔRMŎT'RŌPĬSM, the property or phenomenon of movement under the influence of heat or cold. Curvature toward a source of heat is called *positive thermotropism*; curvature in the opposite direction, *negative thermotropism*.

THĬCK'ENĬNG LĀY'ĔR, an apparent layer of cellulose deposited upon the *inner* surface of a cell-wall. It appears as a layer only because of its power of absorbing water in a different degree from the remainder of the cell-wall.

THĬCK'ENĬNG RĬNG, the cambium-ring, or any other ring or layer of thickening or meristematic tissue.

THŌRN, a degenerated, sharp-pointed branch, either simple, as in the thorn-apple, or branched, as in the honey-locust. Sometimes applied to other large, sharp, rigid processes. Compare SPINE and PRICKLE.

THREAD-SHAPED, see FILIFORM.

THREE-AN'GLED, see TRIGONOUS.

THREE-CLEFT, see TRIFID.

THREE-LEAVED, see TRIFOLIATE.

THREE-LOBED, having three lobes or segments.

THREE-NERVED, having three principal veins proceeding from the base of the leaf in monocotyledons. Applied mainly in the floral envelopes of grasses; trinervate; triple-nerved.

THREE-PART'ED, divided into three parts, or having the parts in threes: a three-parted leaf has three lobes or leaflets; a three-parted flower has three petals.

THREE-RANKED, in three vertical rows upon a stem; trifarions; tristichous.

THREE-VALVED, having three valves or dehiscent portions of a pericarp; trivalvular.

THROAT, see FAUX.

THRUM, an old term for stamen.

THRUM-EYED, a florist's term for flowers having long, conspicuous stamens and a short style. Compare PIN-EYED.

THYLL, see TYLOSIS.

THY'LOSE, see TYLOSIS.

THYRSE, a compact panicle, like that of the lilac and grape.

THYR'SOID, thyrse-like.

THYR'SUS (pl. **Thyr'si**), see THYRSE.

TIGE [pro. **tēzh**], stem. (Rare.)

TIGELLE', see TIGELLUM.

TIGEL'LUM (pl. **Tigel'la**), see CAULICLE.

TIL'LER, n., a sucker or branch from the base of the stem. (Obs.)

TIL'LER, v. i., to put forth new shoots from the root or around the base of the original stalk, as wheat; stool. Applied mainly to the smaller cultivated plants of the grass family. Compare SUCKER.

TIL'LOW (obs.), see TILLER.

TIM'BER-LINE, the upper limit of arborescent vegetation upon high mountains, as determined by cold.

TINCTO'RIOUS, capable of serving as a dye.

TIS'SUE, a general term for all the material of which plants and animals are formed; particularly, a collection of cells of similar character, as vegetable tissue, epidermal tissue, fibrovascular tissue.

TIS'SUE-CORD, see CENTRAL CORD.

TO'MENTOSE, covered with matted woolly hairs.

TOMEN'TOUS, see TOMENTOSE.

TOMEN'TULOSE, slightly tomentose.

TOMEN'TUM, matted woolly hairs.

TOMIP'AROUS, producing spores by division. (M. C. Cooke.) See FISSIPAROUS.

TONGUE, see LIGULE.

TONGUE-SHAPED, long, thickened, nearly flat, and rounded at the end; lingulate; linguiform; linguæform.

TON'OPLAST, a term sometimes applied to a vacuole-wall. See VACUOLE.

TOOTH, any small, pointed, marginal lobe, especially of a leaf.

TOOTHED, see DENTATE.

TOOTH'LET, a small or secondary tooth; denticulation.

TOOTH'LĔTĔD, see DENTICULATE.

TŎP'ĬCAL, local; confined to a limited area.

TŎP'-SHĀPED, see TURBINATE.

TŌRN, with marginal incisions deep and irregular. (Rare.) Compare JAGGED, LACINIATE, and INCISED.

TŌRŌSE', cylindrical, and swollen at intervals. See MONILIFORM.

TŌR'SION, the state of being twisted spirally.

TŌR'TŪOŬS, bending or turning in various directions.

TŌR'ŬLŌSE, somewhat torose.

TŌ'RŬS, (1) the extremity of the stem, upon which the floral organs are situated, usually termed Receptacle; thalamus; (2) a thickened centre in the closing membrane of a bordered pit.

TRÀBĔC'ŬLÀ (pl. **Tràbĕc'ūlæ**), one of the transverse processes upon the inner face of the teeth of the peristome in mosses; a projection from the wall across a cavity, as the bands of tissue crossing the sporangia of Ĭsoetes or the cellular filaments across the space surrounding the fibrovascular bundles in Selaginella; any cross-bar, as one of the connecting threads in a reticulum.

TRÀBĔC'ŬLAR, having or pertaining to trabeculæ; trabeculate.

TRÀBĔC'ŬLAR VĔS'SĔL, one whose cavity is crossed by ligneous threads or bands.

TRÀBĔC'ŬLĀTE, having trabeculæ; trabecular.

TRĀ'CHĔÀ (pl. **Trā'chēæ**), see SPIRAL VESSEL.

TRĀ'CHĔĪD (pl. **Trā'chēids** or **Tràchē'idēs**), a woody vessel composed of a single cell. Applied mainly to those having bordered pits.

TRĀIL'ĬNG, elongated, and prostrate upon the ground, but not rooting. The same as Running, except that the plant may be shorter, or may rise at first from a woody or otherwise self-supporting base. Compare CREEPING.

TRÀJĔC'TĪLE (obs.), see DISTRACTILE.

TRĀ'MÀ, the substance of the same character, as the pileus in the gills of agarics. Ĭt supports the subhymeneal layer.

TRĂNSFŎRMĀ'TION. This term is used in botany mainly in the same sense as Metamorphosis. A petal, for example, is called a transformed or metamorphosed leaf, not because the particular petal under consid cration has ever been a foliage-leaf, but because it is a leaf in a special or unusual condition. The word Transformation is also used for the changes in a particular organ during its development or growth.

TRĂNSMŪTĀ'TION, see METABOLISM.

TRĂNSPĪRĀ'TION, the normal escape of fluids from within the plant; exhalation. Ĭt includes not only the escape of moisture, both in the form of vapor and, as occasionally occurs, in the liquid state, but also the gaseous products of respiration. The transpiration of moisture is sometimes, but unnecessarily, termed Perspiration.

TRĂNSVĔRSE' CHŌ'RĬSĬS, when two or more organs in place of one stand one above or within another; vertical chorisis.

Compare COLLATERAL CHORISIS.

TRĂNSVĔRSE′ ĠĒŎT′RŌPĬȘM, see DIAGEOTROPISM.

TRĂNSVĔRSE′ HĒLĬŎT′RŌPĬȘM, see DIAHELIOTROPISM.

TRȦPĔ′ZĬFŌRM, unsymmetrically four-sided, like a trapezium; trapezoid.

TRĂP′ĔZOID, see TRAPEZIFORM.

TRĂP′ĔZŌĬD, n., a body having the form of a trapezium. (Rare.)

TREE, a woody plant capable of growing in the given locality at least twenty feet in height, with a single self-supporting stem. If the top is unusually broad the height may be somewhat less. The standard is arbitrary and varies considerably according to circumstances. Compare SHRUB.

TRĔM′ĔLLOID, gelatinous, like the Tremellineæ.

TRI-, three.

TRĪȦDĔL′PHOŬS, having the filaments in three sets.

TRĬĂN′DRĬAN, see TRIANDROUS.

TRĬĂN′DROŬS, having three stamens.

TRĬĂN̲′GŬLAR, having three sides or angles;—applied either to flat bodies like leaves, or to columnar bodies like stems. Compare TRIGONOUS.

TRĬĂN′THOŬS, three-flowered.

TRĪ′ÄRCH, a fibrovascular cylinder which has three ligneous rays, and thus represents three fibrovascular bundles.

TRĪBE, a group of genera subordinate to an order or suborder. Similar groups are in some cases called families.

TRĬCÄR′PĔLLĀRY̆, having three carpels in a flower or pistil; trigynous.

TRĬCÄR′POŬS, containing three ovaries or fruits.

TRĪÇĔPH′ȦLOŬS, three-headed.

TRĬCH′ŌBLĂST, an internal hair, like those which project into the intercellular spaces of some water-lilies.

TRĬCH′ŌĠY̆NE, a slender prolongation of the carpogonium in Florideæ which receives upon its apex the antherozoids in fertilization.

TRĬCHŌ′MȦ (pl. **Trĭchō′mȧtȧ**), a term sometimes applied to filaments of conferva and some other algæ, especially to the multicellular filaments in Nostochineæ.

TRĬCH′ŌME, a plant-hair of any kind.

TRĬCH′ŌPHŌRE, the cell or cells immediately supporting a trichogyne.

TRĬCHŌSPŌRĂN′ĠĬŬM, a sporangium which is morphologically a hair. Formerly also applied to the plurilocular sporangia of Phæosporeæ.

TRĬCHŎT′ŌMOŬS, dividing at the end into three branches; three-forked; trifurcate.

TRĬCŎC′COŬS, of three cocci.

TRĪ′CŌLOR, of three colors.

TRĬCŌS′TĀTE, three-ribbed.

TRĬCŬS′PĬD, three-pointed; tricuspidate.

TRĬCŬS′PĬDĀTE, see TRICUSPID.

TRĬDĔN′TĀTE, three-toothed.

TRĬĔN′NĬAL, a plant which fruits the third year, then dies. Few plants, if any, are strictly of this character. Compare BIENNIAL.

TRĬFĀ′RĬOŬS, in three rows, or pointing in three directions; especially, in three vertical rows upon a stem; three-ranked; tristichous. Compare TRISERIAL and MULTIFARIOUS.

TRĬ'FĬD, three-cleft; divided into three parts about half-way to the base.

TRĬFLŌ'ROŬS, three-flowered.

TRĬFŌ'LĬATE, strictly, three-leaved, but often incorrectly applied to leaves which have three leaflets, as those of clover. Compare TRIPHYLLOUS.

TRĬFŌ'LĬOLĀTE, having three leaflets. Usually but incorrectly called Trifoliate.

TRĬFŬR'CĀTE, see TRICHOTOMOUS.

TRĬG'ĀMOŬS, having staminate, pistillate, and perfect flowers in the same head in Compositæ, being one of the polygamous conditions.

TRĪĠĔM'ĬNOŬS, see TERGEMINATE and TRIJUGATE.

TRĬG'ŌNAL, see TRIGONOUS.

TRĬG'ŌNOŬS, prismatic, and three-angled, as the stems of sedges; trigonal. Compare TRIQUETROUS.

TRĬĠ'ȲNOŬS, having three pistils or styles; tricarpellary.

TRĬJ'ŬGĀTE, having three pairs of leaflets arranged in any manner upon a leaf, especially along a common petiole; trijugous. Compare TERGEMINATE.

TRĬJ'ŬGOŬS, see TRIJUGATE.

TRĪLĂT'ĔRAL, three-sided. See TRIGONOUS.

TRĪLŌ'BĀTE, three-lobed.

TRĪLŎC'ŪLAR, three-celled: applied to pericarps.

TRĬ'MĔROŬS, having the parts in threes: applied to flowers having calyx, petals, and stamens each three or a multiple of three. Sometimes written 3-merous.

TRĬMŌR'PHĬC, see TRIMORPHOUS.

TRĪMÔR'PHOŬS, having three kinds of flowers in the same species, differing in the relative lengths of their stamens and pistils; heterogonous trimorphous. Compare DIMORPHOUS. See HETEROGONOUS.

TRĬNĔRV'ĀTE, see THREE-NERVED.

TRĬ'NĔRVED, see THREE-NERVED.

TRĬNŌD'AL, having three nodes.

TRĪŒ'ÇIOŬS, having a polygamous condition in which there are staminate, pistillate, and perfect flowers, each on different sets of plants.

TRĬŌĪ'COŬS, see TRIŒCIOUS.

TRĪŌ'VŪLATE, containing three ovules.

TRĬ'PÄRTĔD, see TRIPARTITE. Compare THREE-PARTED.

TRĬP'ÄRTĪTE, divided into three parts nearly to the base; three-parted.

TRĪPĔT'ALOŬS, of three petals.

TRĬPH'ȲLLOŬS, having three leaves or leaf-like bodies in a whorl, or otherwise associated; —applied especially to a three-leaved calyx or perianth.

TRĪPĬN'NĀTE, thrice pinnate, as a bipinnate leaf the leaflets of which (or some of them) are themselves pinnate; terpinnate.

TRĪPĬNNĂT'ĬFĬD, thrice pinnatifid; thrice pinnately cleft; —said of a pinnatifid leaf when its segments are pinnatifid and the subdivisions of these are also pinnatifid.

TRĪPĬNNĂT'ĬSĔCT, the same as Tripinnatifid, with the divisions extending in each case to the base or midrib.

TRĬP'LE-NĔRVED, see THREE-NERVED.

TRĬP'LE-RĬBBED, having three prominent ribs or veins.

TRĬP'LĬCĀTE-GĔM'ĬNĀTE, see TERGEMINATE.

TRĬP'LĬCĀTE-PĬN'NĀTE, see TRIPINNATE.

TRĬP'LĬCĀTE-TĔR'NĀTE, see TRITERNATE.

TRĬP'LĬNĔRVED, see THREE-NERVED.

TRĬP'TĔROŬS, three-winged.

TRĬQUĔ'TROŬS, prismatic, with three acute or salient angles. Compare TRIGONOUS.

TRĬSĔCT'ĔD, divided into three parts or segments by incisions extending to the midrib or base: said of leaves.

TRĬSĔP'ALOŬS, of three sepals; triphyllous.

TRĬSĔP'TĀTE, having three septa.

TRĬSĒ'RĬAL, in three rows; triseriate. See THREE-RANKED.

TRĬSĒ'RĬĀTE, see TRISERIAL.

TRĬSPĔR'MOŬS, three-seeded.

TRĬS'TĬCHOŬS, see THREE-RANKED.

TRĬSTĬGMĂT'ĬC, having three stigmas.

TRĬS'TĬS, of some dull or dingy color. (Rare.)

TRĬSTȲ'LOŬS, having three styles.

TRĬSŬL'CĀTE, three-grooved.

TRĬTĔR'NĀTE, thrice ternate, as a compound leaf whose primary petiole divides into three secondary petioles, each of which again divides into three, each division bearing three leaflets.

TRĬVĂL'VŬLAR, see THREE-VALVED.

TRĬV'ĬAL NĀME, see SPECIFIC NAME.

TRŌCH'LĔAR, short-cylindrical, with the sides contracted; pulley-shaped; trochleariform.

TRŌCHLĔĂR'ĬFŌRM, see TROCHLEAR.

TRŌPH'ŌPLĂST, a term including all essential granules in protoplasm.

TRŌPH'ŌSPĔRM, see PLACENTA.

TRŬMP'ĔT-SHĀPED, tubular, long, very gradually expanded toward the summit, and having a comparatively small and usually but slightly spreading limb or border. Compare HYPOCRATERIFORM.

TRŬN'CĀTE, terminating abruptly, as though cut off or flattened at the end; extremely obtuse. Compare PREMORSE and SUCCISE.

TRŬNK, the stem of a tree.

TRŬSS, a popular name for a rather compact, moderate-sized, terminal flower-cluster of any kind, as an umbel, corymb, spike, or raceme.

TRȲ'MȦ, a drupaceous fruit, like the walnut or hickory-nut. It is distinguished from a drupe by being derived from an inferior instead of a superior ovary.

TŪBE, the united portion of a gamopetalous corolla, gamosepalous calyx, or monadelphous andrœcium; any elongated hollow part or organ.

TŪBE'-FŌRM, see TUBE-SHAPED.

TŪ'BĔR, a short, thickened portion of a subterranean branch.

TŪ'BĔRCLE, any small, wart-like excrescence, as those upon the rootlets of various Leguminoseæ.

TŬBĔR'CŬLAR, in the form of a tubercle or having tubercles.

TŬBĔR'CŬLĀTE, having tubercles; tuberculose; tubercular.

TŬBĔR'CŬLĀTĔD, see TUBERCULATE.

TŬ'BĔRCŬLE, a tuberous root, as in the dahlia. (Obs.)

TŬBĔR'CŬLŌSE, consisting of tubercles; bearing many tubercles; tuberculate.

TŬBĔR'CŬLOŬS, see TUBERCULOSE.

TŬBĔRĬF'ĔROŬS, bearing tubers.

TŬ'BĔRŌSE, see TUBEROUS.

TŬ'BĔROŬS, bearing tubers, as a tuberous plant; resembling a tuber, as tuberous roots.

TŬBE'-SHĀPED, tubular and rather long and wide; about the same as Trumpet-shaped, but may be shorter; tubiform; tubæform; tubeform; tubate. See TUBULAR.

TŬ'BŬLAR, in the form of a tube or pipe; fistular; tubulose. Applied in Compositæ to disk-florets, in distinction from the ligulate florets of the ray; also to any gamopetalous or gamosepalous flower, especially if the tube is rather long and wide. See TUBE-SHAPED.

TŬ'BŬLAR FLŌ'RĔT, a disk-flower in Compositæ, when, as is usual, it differs from those of the ray in having a small and regular corolla; disk-floret; disk-flower. Compare LIGULATE FLORET.

TŬ'BŬLĬ, pl., see TUBULUS.

TŪBŬLĬFLŌ'ROŬS, applied to a head of flowers in Compositæ which bears tubular florets only.

TŬ'BŬLŌSE, see TUBULAR.

TŬ'BŬLŬS (pl. **Tū'būlī**), (1) one of the cells surrounding the central siphon in Chara; (2) applied by some to the neck in Pyrenomycetes.

TŬFT'ĔD, see CESPITOSE.

TŪMĔS'ÇENT, slightly tumid.

TŪ'MĬD, somewhat turgid or inflated.

TŬ'NĬC, any integument or investing layer, as a seed-coat, or a peridium; especially, a membranous scale of a bulb, or a dry sheath surrounding one of the lower internodes in certain grasses.

TŬ'NĬCĀTE, covered with a thin, separable coat or tunic.

TŬ'NĬCĀTĔD, see TUNICATE.

TŬ'NĬCĀTĔD BŬLB, a bulb with broad, thin scales which form successive overlapping coats, as in the onion; coated bulb. Compare SCALY BULB.

TŬR'BĬNĀTE, top-shaped; inversely conical, as the pericarp of water-lilies.

TÛRĠĔS'ÇENT, swelling; slightly turgid.

TŬR'ĠĬD, (1) thickened as if swollen, like a tuber; (2) distended with liquid, but not with air. Compare INFLATED and TUMID.

TŬ'RĬŌ (pl. **Tūriō'nēs**), see TURION.

TŬ'RĬŌN, a leafless or scaly shoot from the ground, as a young stem of asparagus; turio.

TŬRĬŌNĬF'ĔROŬS, bearing turions.

TŬRN'ĬNG ĬN, commencing to head: said of cabbages, etc.

TŬR'NĬP-SHĀPED, see NAPIFORM.

TŬR'PĔNTĪNE VĔS'SĔLS, see RESIN-PASSAGES.

TŬS'SŎCK, a tuft growing from the ground, as in many grasses.

TWĬN, see GEMINATE.

TWĬN'ĬNG, ascending by coiling the stem in a spiral manner around a support; voluble. Compare CLIMBING.

TWĬST'ĔD, see CONTORTED.

TWO'-CLĔFT, see BIFID.

TWO'-FŌRKED, see DICHOTOMOUS.

TWO'-LĬPPED, see LABIATE.

TWO'-PÄRTĔD, see BIPARTITE.

TWO'-RĂNKED, situated in two vertical rows on opposite sides of the stem, as the leaves of grasses; distichous.

TWO'-TOOTHED, see BIDENTATE.

TȲ'LŌSE, see TYLOSIS.

TȲLŌ'SĬS (pl. **Tylō'sēş**), a protrusion from an adjoining cell into the cavity of a vessel, sometimes exhibiting repeated cell-division within the vessel. Sometimes written Thylose and Thyll. A vessel which contains these protrusions is said to exhibit tylosis.

TȲM'PĂNŬM, a membrane closing the mouth of the capsule in some mosses.

TȲPE, a perfect specimen or individual, exemplifying the essential characters of the species or other group to which it belongs.

TȲPE-SPĔÇ'IMEN, the original specimen from which a botanical description was written and upon which the name of the plant or group is based: the actual individual which serves as the type of a species or other group.

TȲP'ĬCAL, representing the type or plan.

ŪLĬG'ĬNŌSE, see PALUSTRINE.

ŪLĬG'ĬNOŬS, see PALUSTRINE.

ŬM'BĔL, the inflorescence of the order Umbelliferæ, or any flower-cluster in which several primary rays or pedicels arise from the apex of the stem or peduncle. A typical umbel is somewhat flat-topped, and has the rays spreading like the stays of an umbrella. An umbel is *simple* when each ray or pedicel bears but a single flower; it is *compound* when each bears several flowers or secondary rays.

ŬM'BĔL, ÇȲ'MŌSE, see CYMOSE UMBEL.

ŬM'BĔLLĀTE, bearing umbels; pertaining to umbels; umbel-like.

ŬM'BĔLLĀTE ÇȲME, see CYMOSE UMBEL.

ŬM'BĔLLĔT, a small umbel or a partial umbel; umbellule.

ŬMBĔLLĬF'ĔROŬS, producing umbels.

ŬMBĔL'LĬFŌRM, in the shape of an umbel.

ŬMBĔL'LŬLĀTE, in the form of a small or a partial umbel. Formerly used for Subumbellate, which see.

ŬM'BĔLLŬLE, see UMBELLET.

ŬMBĔLLŬLĬF'ĔROŬS, bearing small umbels.

ŬMBĬL'ĬCAL CŌRD, see FUNICULUS.

ŬMBĬL'ĬCĀTE, having an umbilicus, or central depression like the navel.

ŬMBĬLĬ'CŬS, an old term for Hilum; any depression resembling the navel.

ŬM'BŌ, a central elevation, like the boss of an ancient buckler.

ŬM'BŌNĀTE, having a low, rounded central projection, as the cap of many mushrooms.

ŬMBŌN'ŬLĀTE, slightly umbonate; subumbonate; bearing a small, or slightly elevated, umbo.

ŬMBRĂC'ŬLĬFŌRM, umbrella-shaped.

ŬMBRĂC'ŬLŬM, the fruit-cap of Marchantia, or any such umbrella-shaped appendage.

ŬNÄRMED', destitute of thorns, spines, or prickles.

ŬN'CĀTE, see UNCINATE.

ŬN'ÇĬFÔRM, hook-shaped.

ŬN'ÇĬNĀTE, hooked at the end, or furnished with hooked appendages; uncate; unciform. Compare FALCATE.

ŬNC'TŪOŬS, having a greasy appearance.

ŬN'DĀTE, see UNDULATE.

ŬN'DATED, see UNDULATE.

ŬN'DĔRSHRŬB, (1) a low shrub, less than three feet high, as the wintergreen; (2) a plant with woody base and upper portion herbaceous and yearly dying back, as the garden sage (suffrutex). The term is now generally used in the first sense only.

ŬN'DŪLĀTE, applied to leaves which have the surface near the margin alternately concave and convex; undate. Compare CRISP, SINUATE, WAVED, and RUFFLED.

ŬNĔ'QUAL, (1) unsymmetrical, as the leaves of begonia; (2) differing in length—applied to stamens, etc.

ŬNĔ'QUALLY̆ PĬN'NĀTE, see IMPARIPINNATE.

ŬNĔ'QUAL-SĪDĔD, unsymmetrical.

ŬNGUĬC'ŬLAR, see UNGUICULATE.

ŬNGUĬC'ŬLĀTE, (1) furnished with a "claw" or unguis, as the petals of pinks; (2) ending in a curved point resembling a claw.

ŬN'GUĬFŌRM, like the claw of a petal.

ŬN'GUĬS, see CLAW.

ŬN'GŪLĀTE, hoof-shaped.

UNI-, a Latin prefix, one.

ŪNĬĂX'ĬAL, having an unbranched stem.

ŪNĬCĂP'SŪLAR, having the carpels of a flower all united into one capsule.

ŪNĬCĂR'ĬNĀTĔD, one-keeled.

ŪNĬÇĔL'LŪLAR, one-celled.

ŪNĬCŌL'OR, of the same color throughout; whole-colored; unicolorous; isochrons. Compare DISCOLOR and CONCOLOR.

ŪNĬCŌL'OROŬS, see UNICOLOR.

ŪNĬFĀ'RĬOŬS, one-ranked. Compare SECUND.

ŪNĬFLŌ'ROŬS, one-flowered.

ŪNĬFŌ'LĬĀTE, one-leaved. Used also for Unifoliolate, which see.

ŪNĬFŌ'LĬŌLĀTE, of one leaflet, as the theoretically compound leaf of the orange and lemon.

ŪNĬJ'ŪGĀTE, having a single pair of leaflets or other organs.

ŪNĬLĀ'BĬĀTE, one-lipped;—sometimes applied also to a regular gamopetalous corolla which is open on one side, as the ligulate florets in Compositæ. See LABIATE.

ŪNĬLĂT'ĔRAL, one-sided. See SECUND.

ŪNĬLŎC'ŪLAR, one-celled, as applied to anthers and ovaries; elocular.

ŬNĬNTĔRRŬP'TĔD, see CONTINUOUS.

ŪNĬP'ĂROŬS, bearing or producing but one stem or axis.

ŬNĬP'ÅROŬS ÇȲME, a cyme with one main axis; monochasium. Compare DICHASIUM.

ŪNĬPĔT'ALOŬS, having but one petal, as Amorpha. Compare GAMOPETALOUS.

ŪNĬSĔP'TĀTE, having a single septum.

ŪNĬSĒ'RĬAL, having one row or whorl; uniseriate. Compare ONE-RANKED.

ŪNĬSĒ'RĬĀTE, arranged in a single line; uniserial.

ŪNĬSĔX'ŪAL, applied to an individual or flower which has one kind of sexual organs only; diclinous.

ŪNĬVĂL'VŪLAR, dehiscing along one suture only, so that the pericarp has but one valve, as the pod of the common milkweed, *Asclepias Cornuti.*

ŪNĬVĔR'SAL, see COMMON.

ŪNĬVĔR'SAL ĬN'VŌLŬCRĔ, see COMMON INVOLUCRE.

ŪNĬVĔRSAL ŬM'BĔL, see COMPOUND UMBEL.

ŬNLĬN'ĬNG, the separation of parts originally united. Formerly used for Chorisis, from the erroneous supposition that the additional organs in chorisis were always produced in this manner. See CHORISIS.

ŬNSȲMMĔT'RĬCAL, not symmetrical, which see.

ÛR'ÇĒŌLĀTE, pitcher- or urn-shaped: tubular, and contracted at the orifice.

ŪRĒ'DŌ-FRUIT, a sorus or group of uredospores.

ŪRĒ'DŌSPŌRE, a form of unicellular spore or gonidium in the Uredineæ or rust-fungi, produced earlier in the season than the teleutospores, and destined for immediate germination.

ŪRĒ'DŌ-STĀĠE, the early summer stage of the Uredineæ, during which only uredospores are produced.

ÛRN, the spore-capsule of mosses; also the base of a pyxidium.

ÛRN-SHĀPED, see URCEOLATE.

ÛRTĬCĀ'ÇEOUS, pertaining to nettles or the family Urticaceæ.

Ū'TRĬCLE, (1) a fruit having a small inflated membranous pericarp, as that of Chenopodium; (2) the bladder of various aquatic plants, as Utricularia; (3) one of the large hyaline cells in the leaves of Sphagnum.

Ū'TRĬCLE, PRĬMÔR'DĬAL, see PRIMORDIAL UTRICLE.

ŪTRĬC'ŪLAR, bladder-like or furnished with utricles.

ŪTRĬC'ŪLĀTE, inflated like a bladder; utricular.

ŪTRĬC'ŪLĬFŌRM, shaped like a bottle or bladder; about the same as Urceolate, but a less definite term.

ŪTRĬC'ŪLŌSE, bearing utricles.

ŪTRĬC'ŪLŬS (pl. **Ūtrĭc'ūlī**), see UTRICLE.

VĂC'ŪŌLE, a sap-cavity in the protoplasm of a cell.

VĀĠĪ'NÅ, see SHEATH.

VĂĠ'ĬNANT, sheathing.

VĂĠ'ĬNĀTĔD, sheathed; invaginated.

VĂĠĬNĔRVŌSE', having the small veins (nerves) in no apparent order.

VĀĠĬN'ŪLÅ, a small sheath; especially, the apex of the stem which surrounds the base of the seta in mosses; vaginule.

VĂĠ'ĬNŪLE, see VAGINULA.

VĀGUE, in no definite order or direction; of no definite or constant form. Compare AMORPHOUS.

VĀIL, see VEIL.

VĂLLĔC'ŪLȦ (pl. **Vällĕc'ūlæ**), a groove or furrow, as those between the ridges on the fruit of Umbelliferæ. Compare SULCUS.

VȦLLĔC'ŪLAR CȦNĂLṢ', large intercellular passages, alternating with the fibrovascular bundles in the stem of Equisetum. They are situated in the cortex, and lie between the ridges on the surface. Compare CARINAL CANAL.

VĂLV'ĀTE, said of the leaves of a flower in æstivation when they meet at their edges and do not overlap, as the sepals in the rose; valvular. The margins of valvate organs may be rolled or folded inward or outward.

VĂLVE, (1) one of the parts of a dehiscent pericarp; (2) the lid of an ascidium or pitcher; (3) one of the halves of a frustule in diatoms.

VĂLV'ŪLAR, see VALVATE.

VĀ'RĬȦBLE, said of a species or other group which embraces many individuals which depart more or less from the type of the group. The term is applied in a similar way to any organ or character which fails to exhibit uniformity.

VĀRĬĀ'TION, a transient variety, consisting of but one or a few individuals, less marked than a Sport, and usually but slightly differing from the type of the species or variety to which it belongs. It is subordinate in importance to a Form, and less frequently produced by some peculiar condition of the soil or climate.

VĂR'ĬCŌSE, appearing abnormally enlarged in places;—applied to hairs and other tubular filaments.

VĀ'RĬĔGĀTĔD, applied to leaves, etc., which have two or more colors upon the surface, especially to such as are permanently marked with white or yellow. Compare BICOLOR and CHLOROSIS.

VȦRĪ'ĔTȲ, a group subordinate to a species, founded on characters which in cultivated plants are often temporary, and which in the wild state are regarded as permanent, but which gradually merge into those of the main specific form. Compare SPECIES, RACE, SPORT, and VARIATION.

VȦRĪ'ĔTȲ-HȲ'BRĬD, see CROSS.

VÄR'NĬSH, see BLASTOCOLLA.

VÄR'NĬSHED, see VERNICOSE.

VĀ'SȦ PRŌ'PRĬȦ, a term applied by Mohl to the portion of the phloëm containing the sieve-tubes and other thin-walled tubular cells. Now seldom used.

VĂS'CŪLAR, pertaining to or containing vessels.

VĂS'CŪLAR BŬN'DLE, see FIBROVASCULAR BUNDLE.

VĂS'CŪLAR BŬN'DLE-SHĒATH, a layer of cells between the phloëm and cortex, or surrounding a fibrovascular bundle, or an entire fibrovascular cylinder; phloëm-sheath.

VĂS'CŪLAR ÇȲL'ĬNDĔR, see FIBROVASCULAR CYLINDER.

VĂS'CŪLAR SȲS'TĔM, see FIBROVASCULAR SYSTEM.

VĂS'CŪLAR TĬS'SŪE, see VESSEL.

VĂS'CŪLŬM, a botanist's collecting-case. Formerly this term was applied to an ascidium or pitcher-shaped leaf.

VĀSE'-SHĀPED, shaped somewhat like a common flower-pot. (Rare.)

VĂS'ĬFŌRM, having the character of ducts or vessels.

VĂS'ĬFŌRM WO͝OD-ÇĔLLS, see TRACHEIDS.

VAULT'ĔD, overarched, as the upper lip of many ringent flowers; fornicate.

VĔĠ'ĔTȦBLE, (1) any plant; (2) in horticulture applied to plants cultivated for some edible part besides the fruit, and also to some, as the melon family, in which the part used is properly a fruit.

VĔĠ'ĔTȦBLE ȦNĂT'ŌMY̆, see STRUCTURAL BOTANY.

VĔĠ'ĔTȦBLE NŌSŌL'ŌĠY̆, the department of vegetable pathology which treats of the diagnosis or classification of the diseases of plants.

VĔĠ'ĔTȦBLE PȦTHŎL'ŌĠY̆, the science which treats of the diseases of plants. It includes the study of disease-producing parasites and of the means for preventing their injuries. It does not include Teratology.

VĔĠ'ĔTȦBLE PHY̆ŞĬŎL'ŌĠY̆, see PHYSIOLOGICAL BOTANY.

VĔĠ'ĔTȦBLE TĂXŎN'ŌMY̆, the part of systematic botany which relates to the classification of plants.

VĔĠ'ĔTȦBLE WĂX, a wax-like substance upon many leaves and fruits, an important function of which is to protect the parts from excessive moisture. See BLOOM.

VĔĠĔTĀ'TION, (1) germination and growth; (2) plants in general.

VĔĠ'ĔTĀTĬVE ȦPŌG'ȦMY̆, the apogamous production of growing shoots in place of seeds or spores. See APOGAMY.

VĔĠ'ĔTĀTĬVE ÇĔLL, a cell in a pollen-grain which does not develop into the pollen-tube. Compare GENERATIVE CELL.

VĔĠ'ĔTĀTĬVE NŬ'CLĔŬS, any nucleus in a pollen-tube which takes no direct part in fertilization. Compare GENERATIVE NUCLEUS.

VEIL, (1) a membrane connecting the margin of the cap in mushrooms with the stalk; velum; (2) the calyptra in mosses.

VEIN, a small bundle of fibro-vascular tissue within a leaf. When large, and occupying a prominent ridge, it is called *rib;* when very small it is sometimes called *nerve,* especially in monocotyledons; but the term Nerve is now little used.

VEIN'LĔT, a small, secondary vein; veinule.

VEIN'ŪLĔT, a branch of a veinlet. (Rare.)

VĔLĀ'MĔN, an envelope of several layers of cells containing air, surrounding the aërial roots of orchids and Aroideæ.

VĒ'LĀTE, furnished with a veil; veiled.

VĒ'LŬM, the membranous indusium in Isoetes. Formerly applied to the veil of mushrooms.

VĒ'LŬM PÄRTĬĀ'LĔ, see MARGINAL VEIL.

VĒ'LŬM ŪNĬVĔRSĀ'LĔ, see VOLVA.

VĔLŪ'TĬNOŬS, covered with a close, silky coat of short, fine, erect hairs of even length; velvety.

VĔL'VĔTY̆, see VELUTINOUS.

VĔNĀ'TION, the manner in which the veins are arranged in a leaf.

VĔNŌSE', containing numerous veins.

VĔN'TĔR, the expanded basal portion of an archegonium, in which the oösphere is formed.

VĔN'TRAL, pertaining to the face or front, being the side opposite to the dorsal. See DORSAL.

VĔN'TRAL CȦNĂL' ÇĔLL, a small cell below the entrance of the neck of an archegonium, cut off from the mother-cell of the oösphere.

VĔN'TRAL SŬ'TŪRE, a line of union between the margins of the carpel or carpels in an ovary. Compare DORSAL SUTURE.

VĔN'TRĬCŌSE, swelling out in a rounded manner.

VĔNTRĬC'ŪLŌSE, slightly ventricose.

VĔNTRĬCŬM'BENT, face downward upon the ground; prone.

VĔN'ŪLŌSE, abounding in veinlets.

VĔRMĬC'ŪLAR, worm-shaped.

VĔRMĬC'ŪLĀTE, vermicular, or bearing worm-like processes.

VĔR'MĬFŌRM BŌD'Ў, see SCOLECITE.

VĔR'NAL, pertaining to spring; appearing in spring.

VĔRNĀ'TION, the manner in which leaves are disposed in the bud; prefoliation. Sometimes improperly used for Foliation, the act or time of leafing.

VĔR'NĬCŌSE, appearing as though varnished, as the buds of many trees.

VĔRRỤ'CȦ (pl. **Vĕrrụ'çæ**), a wart-like elevation of any kind, including the sessile apothecia of some lichens. (Obs) See WART.

VĔRRỤ'CÆFÔRM, wart-like.

VĔR'RỤCŌSE, wart-like, or bearing wart-like prominences. Compare PAPILLATE and PUSTULATE.

VĔRRỤ'CŪLŌSE, slightly verrucose.

VĔR'SȦTĬLE, attached at one point, so as to swing freely to and fro, as the anthers of the lily.

VĔR'SĬCŌLOR, changeable in color, or appearing of different colors from different points of view.

VĔR'SĬFŌRM, varied or varying in form.

VĔR'TĔX, an upper extremity; summit. Compare APEX.

VĔR'TĬCAL, (1) perpendicular to the horizon; (2) perpendicular to the surface or axis of support; (3) in the direction of the axis of growth; lengthwise.

VĔR'TĬCAL ĂN'THĔR, see INNATE.

VĔR'TĬCAL CHŌ'RĬSĬS, see TRANSVERSE CHORISIS.

VĔR'TĬCAL LĔAVEȘ, applied especially to erect leaves like those of Iris, which have no distinct or obvious dorsal and ventral surfaces.

VĔR'TĬCALLЎ CŎMPRĔSSED', see DEPRESSED.

VĔR'TĬCAL SЎS'TĔM, see FIBRO-VASCULAR SYSTEM. Compare HORIZONTAL SYSTEM.

VĔR'TĬÇĔL, see VERTICIL.

VĔR'TĬÇĬL, see WHORL.

VĔRTĬÇĬLLĂS'TĔR, a pair of dense cymes in the axils of opposite leaves, forming an apparent verticil, as in most Labiatæ.

VẼRTĬÇ'ĬLLĀTE, arranged in a ring or whorl; whorled.

VĔS'ĬCLE, a small bladder-like body or cavity.

VĔSĬC'ŬLAR, bearing or containing numerous vesicles; vesiculate; vesiculose.

VĔSĬC'ŬLĀTE, see VESICULAR.

VĔSĬC'ŬLŌSE, see VESICULAR.

VĔS'PĔRTĬNE, pertaining to or taking place early in the evening: applied especially to flowers which expand at that time of the day.

VĔS'SĔL, a cell, or row of confluent cells, having markings produced by thickenings in the cell-wall, forming dots, bands, rings, spirals, etc. Formerly only vessels having spiral markings were called vessels, in distinction especially from those having pitted walls, which were known as ducts, but the term Duct is now comparatively little used. The terms Duct and Vessel are partially equivalent, Duct being perhaps more often used for continuous vessels composed of more than one cell. See TRACHEA and TRACHEID.

VĔS'TĬBŬLE, a term applied by Duchartre to an opening or chamber above the stomata in certain plants, as *Cycas revoluta,* formed by the depression of the guard-cells and the outward growth of the adjoining epidermal cells. Compare ANTECHAMBER and STOMATIC CHAMBER.

VĔS'TĬĠE, an abortive part which in some ancestral form was fully developed.

VĔT'ĔRAN (Forestry), a very old tree.

VĔX'ĬL, see VEXILLUM.

VĔX'ĬLLĀRY̆ ÆSTĬVĀ'TION, having a vexillum or other similar large petal which enfolds all the others in the bud. More properly "Cochlear Æstivation," except in Papilionaceæ.

VĔX'ĬLLĀTE, having a vexillum.

VĔXĬL'LŬM, the large upper petal of a papilionaceous flower; standard; banner.

VĪ'ȦBLE, capable of growing or manifesting life—said mainly of seeds.

VĪ'BRȦTĬLE, having a vibratory or back-and-forth motion.

VĪ'BRŌĠĔN, a term applied by Penhallow to certain longitudinal bands of active tissue in the cortex of tendrils to which their movements of circumnutation are supposed to be chiefly due.

VĬCĀ'RĬOŬS, performing the function, or occupying the place, of some other organ.

VĬL'LĪ (sing. **Vĭl'lŭs**), velutinous hairs.

VĬL'LĬFŌRM, resembling villi.

VĬLLŌSE', covered with rather thin, soft, straightish hairs, somewhat finer, softer, and thicker than in Pilose; villous. Sometimes used for Velutinous, which see.

VĬL'LOŬS, see VILLOSE.

VĬL'LŬS, sing., see VILLI.

VĪ'MĔN, a long, flexible shoot or branch, like the willow branches used for wickerwork. (Obs.)

VĬM'ĬNAL, pertaining to twigs.

VĬMĬN'ĔOŬS, producing or resembling long, flexible twigs or branches.

VĪNE, any plant, or its stem, which is incapable of self-support, and either grows upon the ground or supports itself upon other objects.

VĪRĔS'ÇENÇE, having petals abnormally green in color, like leaves.

VĪRĔS'ÇENT, greenish, or becoming green.

VĬR'GĀTE, wand-like; long, straight, and slender.

VĬR'GŬLĀTE, diminutive of Virgate; shaped like a little twig or rod.

VĬRĬDĔS'ÇENT, greenish; virescent.

VĬRŌSE', having a nauseous odor.

VĬS'ÇĬD, sticky; adhesive; viscous.

VĬS'ÇĬD DĬSK, the retinaculum of orchids.

VĬTĔL'LŬS, the embryo-sac as it appears in the seed.

VĬTĬC'ŌLOŬS, growing upon the grape-vine, as certain parasitic fungi.

VĪTĬC'ŪLÀ (pl. Vītĭc'ūlæ), a trailing stem, as that of the cucumber; a little vine; a tendril; a stolon; a sarment. (Rare.)

VĪTĬC'ŪLŌSE, bearing or resembling viticulæ.

VĬT'RĔOŬS, transparent like glass. Compare HYALINE.

VĬT'TÀ (pl. Vĭt'tæ), one of the tubular oil-receptacles in the fruit of Umbelliferæ; oil-tube. They appear externally as ridges.

VĬT'TĀTE, bearing vittæ, or longitudinal stripes or ridges somewhat resembling the vittæ on the fruit of Umbelliferæ.

VĪVĀ'ÇIOŬS, (1) living over winter, or from year to year; perennial; (2) full of life; difficult to kill.

VĪVĬP'ÀROŬS, producing bulbs or seeds which germinate while still attached to the parent-plant. The term is not applied to the production of ordinary offsets or suckers.

VŎL'ŪBĬLE, see TWINING.

VŎL'ŪBLE, see TWINING.

VŎLŬNTEER'. A volunteer crop or plant is one which has come from self-sown seed within the boundaries of the field or farm and has grown without care or cultivation. The term implies that the growth is of some value.

VŌLŪTE', rolled up in any direction.

VŎL'VÀ, a covering, membranous, gelatinous, or otherwise, which extends from the margin of the cap to the base of the stipe in mushrooms, and which is ruptured by growth in those species which have an elongated stem; wrapper; velum universale.

WALL'ĬNG Ō'VĔR, a term employed in forestry for the growth of a callus over a cut or injured surface.

WART, a sessile gland, hard excrescence, or other protuberance resembling a wart upon animals. Compare PUSTULE.

WART'Ў, applied by gardeners to a head of cauliflower when it becomes loose and covered with small prominences preparatory to throwing up the seed-stalks; frothy.

WA'TĔR-CŌRE, an abnormal condition of a part or the whole of the flesh of the apple, occasionally seen, in which the cells are unusually firm, do not lose their fluid contents upon ripening, and appear as though water-soaked.

WA̤'TĔR-GLĂND, a group of cells beneath the water-pores of certain plants, as in Saxifraga, which serve to facilitate or regulate the transpiration of water. The cells composing the gland are rich in protoplasm but destitute of chlorophyll.

WA̤'TĔR-PŌRE, an aperture in the epidermis directly over the extremity of a vein in the leaves of certain plants, as *Caladium esculentum.* It is a stoma whose guard-cells have lost their function and become immovable, and its probable function is to facilitate the transpiration of water.

WA̤'TĔR-SPROUT (Hort.), an unusually long shoot or limb of one year's growth, especially upon the body of a tree where older limbs have been removed.

WA̤'TĔR-STŌMȦ, see WATER-PORE.

WA̤'TĔR-STŌR'ĬNG TĬS'SŬE, a form of water-tissue adapted for the storage of water for the supply of surrounding cells in times of drought, as in many plants of dry climates.

WĀVED, said of a margin when strongly undulate, but less strongly than Ruffled.

WĀV'Y̆, alternately concave and convex upon the surface or margin. See SINUATE, REPAND, UNDULATE, and CRISP.

WĔATH'ĔR-PROUD, a gardener's term, employed in England, denoting that plants are larger or more thrifty than common for the season of the year.

WĔDĠE'-SHĀPED, see CUNEATE.

WEED, a plant which grows where it is not wanted and which becomes troublesome. Applied chiefly to wild plants.

WEEP'ĬNG, (1) having the smaller branches so lax that many of them hang directly downward, as those of several species of willow; (2) the copious exudation of sap from water-pores or from an injured surface, in the latter case more often called Bleeding.

WĔFT, a collection of interlacing hyphæ or other filaments.

WĔLT, a broad, raised stripe or ridge upon the surface of a fruit, as is occasionally seen in the orange and lemon.

WĔLT'ĔD, flaccid; drooping. (Obs)

WHEEL'-SHĀPED, see ROTATE.

WHĬP'-SHĀPED, see FLAGELLIFORM.

WHĬRL, see WHORL.

WHŌRL, a set of organs arranged in a circle around a stem or axis; verticil.

WHÔRLED, see VERTICILLATE.

WĬLD, (1) growing spontaneously without care or cultivation, either native or introduced; feral; (2) in a natural condition, not improved by cultivation or artificial selection. When used without qualification Wild means strictly native.

WĬLD'ĔRĬNG, see WILDING.

WĬLD'ĬNG, (1) a cultivated plant that has "run wild" or reproduced itself spontaneously; wildering; (2) any wild plant; wildling. In both senses applied to fruit-trees. A fruit-tree, however, which has sprung up spontaneously within cultivated ground is not usually called a Wilding, but merely a "chance seedling."

WĬNG, (1) any thin or membranous appendage, especially a flattened appendage to a seed which enables it to be more readily conveyed by the wind, as in the ash and maple; (2) one of the side petals of a papilionaceous flower, as of the pea or bean; ala.

WĬNGED, bearing a membranous expansion or wing; alate.

WĬN'TĔR AN'NŬAL, a plant which germinates in the fall, lives over winter, and produces seed and dies the following season.

WĬN'TĔR SPŌRE, see RESTING-SPORE.

WĬTH'ĔRĬNG, see MARCESCENT.

WOOD, all that portion of the stem in exogens within the bark or cambium layer, except the pith and medullary rays. It consists of the xylem portions of the united fibrovascular bundles. The term is also extended in popular use to all parts of similar texture in plants. See XYLEM.

WOOD-FĪ'BRE, an elongated wood-cell with thick wall and narrow cavity. See BAST-FIBRE.

WOOL, long, dense, curled hairs.

WOOL'LY̆, see LANATE.

WŎR'ŌNĬN'S HY̆'PHȦ, a coiled hypha in various Ascomycetes, believed to be morphologically an archicarp, but in which fertilization is either wanting or unknown.

WRĂP'PĔR, see VOLVA.

XĂN'THĬC FLOW'ĔRS, those which have yellow for the typical color, including all colors except those containing blue. Compare CYANIC FLOWERS.

XĂN'THŌPHY̆LL, see PHYLLOXANTHIN.

XĔNŎG'ȦMY̆, see CROSS-FERTILIZATION.

XĔRŌPH'ĬLOŬS, growing in dry climates. (Rare.)

XĬPH'ĪOID (rare in botany), see ENSIFORM.

XĬPHŌPH'Y̆LLOŬS, having ensiform leaves. (Rare.)

XȲ'LĔM, the portion of a differentiated fibrovascular bundle which contains the larger continuous air-containing vessels, and the walls of whose cells are often thickened and lignified; wood; hadrome. The xylem is separated from another part called Phloëm by the cambium, if there is any, and it usually occupies the side of the bundle toward the centre of the stem. Compare PHLOËM.

XȲ'LĔM-PLĀTE, see XYLEM-RAY.

XȲ'LĔM-RĀY, a radial plate of xylem between two medullary rays; xylem-plate. Compare PHLOËM-RAY.

XY̆LŌCÄR'POŬS, having hard, woody fruit. (Obs.)

XY̆LŌ'DĬŬM (obs.), see ACHENIUM.

XY̆LŌ'MȦ, a sclerotoid body which produces spores internally: found in Polystigma and some other Ascomycetes.

ZŌĂD'ŪLȦ (pl. Zōăd'ūlæ) (obs.), see ZOÖSPORE.

ZŌDĪŌPH'ĬLOŬS, pollinated by the agency of animals, including, for example, Protozophilous and Entomophilous. It is the converse of Anthophilous, said of the insects or other animals which convey the pollen. Compare ANEMOPHILOUS and HYDROPHILOUS.

ZŌN'ĀTE, marked with circular, colored bands; zoned. Applied to the tetraspores of Florideæ when in a single row, in distinction from a cruciate or tripartite arrangement.

ZŌNE, a circular band or stripe.

ZŌNED, see ZONATE.

ZŌ'ŌGÅMĔTE, see PLANOGAMETE.

ZŌŌGLŒ'Å (pl. **Zōōglœ'æ**), a gelatinous colony of bacteria.

ZŌŌGŌNĬD'ĬŬM (pl. **Zōōgōnĭd'ĭå**), a motile gonidium, usually called zoöspore. Restricted by some to a zoöspore which grows into a plant directly, without conjugation. See ZOÖSPORE.

ZŌ'ŌSPĔRM, see ZOÖSPORE.

ZŌŌSPŌRĂN'ĠĬŬM, a sporangium enclosing zoöspores.

ZŌ'ŌSPŌRE, a spore or gonidium having the power of independent movement, usually by means of one or more vibratile cilia; swarm-spore. See ZOÖGONIDIUM.

ZŌŌTHĒ'CÅ (obs.), see ZOÖSPORANGIUM.

ZY̆GŌGŌ'NĬŬM, the female conjugating cell in Conjugateæ, together with the accessory cells, if any. Compare ASCOGONIUM and CARPOGONIUM.

ZY̆GŌMŌR'PHĬC, applied to flowers which are monosymmetrical, as those of the pea and bean. Compare ACTINOMORPHIC. Sachs extends the term to such irregular flowers as are capable of bisection into similar halves in two directions, but in which the halves produced by the two bisections are of different shape, as in Dicentra.

ZY̆GŌMŌR'PHOŬS, see ZYGOMORPHIC.

ZY̆GŌ'SĬS, see CONJUGATION.

ZY̆G'ŌSPĔRM, see ZYGOSPORE.

ZY̆G'ŌSPHĔRE, the female conjugating cell in such zygophytes as have a distinction in sex. Compare OÖSPHERE.

ZY̆G'ŌSPŌRE, a spore resulting from conjugation.

ZY̆'GŌTE, a general term for the product of the union of two gametes.

ZY̆GŌZŌ'ŌSPŌRE, a motile zygospore.

ZȲM'ĀSE, a ferment secreted by certain bacteria; enzyme.

ZȲMŌ'SĬS, (1) fermentation of any kind; (2) an infectious or contagious disease.

ZȲMŌT'ĬC, (1) pertaining to fermentation, or caused by organisms capable of producing fermentation; (2) pertaining to an infectious or contagious disease.

Made in the USA
Monee, IL
05 November 2026

67383972R00125